THE WORLD'S WARSHIPS 1915.

COMPILED FROM JANE'S "FIGHTING SHIPS"

GREAT BRITAIN.

DREADNOUGHTS—*continued*.

		Completed	Displacement (tons)	Speed (knots)	Armament	Complement
1909 Dreadnoughts. *St. Vincent class.*	St. Vincent à Collingwood . Vanguard .	1910	19,250	21	Guns: 10—12 inch, 50 cal.; 18—4 in. 4—3 pdrs. Torpedo tubes (18 inch): 3.	870
1907 Dreadnoughts. *Téméraire Class.*	Bellerophon . Téméraire . Superb .	1909	18,600	20.75	Guns: 10—12 inch, 16—4 inch; 4—3 pdrs. (saluting). Torpedo tubes (18 inch): 3.	850
1906 Dreadnought.	Dreadnought	1906	17,900	21	Guns: 10—12 inch, 45 cal.; 24—12 pdrs. Torpedo tubes (18 inch): 5.	800

5

Plate XIII

H.M.S. "DREADNOUGHT" (Battleship). Cribb

FRED T. JANE

PREFACE TO THE 2022 EDITION BY FRED T. JANE [AI]

NIMBLE BOOKS LLC

Publishing Information:

ISBN: 978-1-60888-010-2 (hc)

Publisher's Note

Receiving a copy of *Jane's Fighting Ships 1969-70* on my 10th birthday was one of the highlights of my childhood, and Jane's has shone brightly in my mind ever since. Ever since its first edition in 1898, *Jane's* has been the go-to source for information for navalists. As a result past issues of *Jane's* are an irreplaceable source of contemporaneous perspectives on the world's navies. Flipping through *Jane's* for 1898 or 1906 or 1939, you are seeing through the eyes of history. Unfortunately, most past editions are no longer in print and can only be obtained by purchasing used copies, which can be expensive and scarce, or via interlibrary loan, which can be slow and unreliable.

The World's Warships were condensed wartime versions of the full editions of *Jane's Fighting Ships* that provided lists of ships by class and name, photographs, and basic particulars such as number of guns, while omitting the detailed technical information found in *Fighting Ships.* The 1915 edition was published in a 5.50 x 4.10" landscape format. Unfortunately, this likably compact format is not available from the large US print-on-demand vendors, so this reprint is published in 11" x 8.5" landscape with page images expanded to fit.

This edition is annotated with a contemporary photograph of Fred T. Jane as a young man; a modern photograph of the "blue plaque" historical marker placed at the site of his residence in Southsea; and a second, modern preface written in 2022 from the point of view of the AI spirit of Fred T. Jane.

I hope you enjoy this edition as much as I enjoyed that first Jane's.

Fred Zimmerman
Nimble Books LLC
Ann Arbor, Michigan, USA

PREFACE TO THE 2022 EDITION.

It is now over 100 years since the first edition of All the World's Fighting Ships was published, and in that time there have been many changes in the naval landscape. In 1898, when Jane's first came out, there were only a few major powers with large navies: Britain, France, Russia, Italy, Austria-Hungary, and the United States. Today, there are many more states with significant naval forces, including China, India, Japan, South Korea, Iran, and Saudi Arabia.

The types of ships in service have also changed drastically. In 1898, the largest warships were battleships, which were armed with heavy guns and designed to engage enemy fleets in direct combat. Today, nuclear-powered aircraft carriers are the most powerful vessels in the world. These ships carry dozens of fighter jets and helicopters, and can project military power around the globe.

The role of navies has also changed over time. In the early 20th century, navies were primarily used for offensive purposes, such as blockading enemy ports or bombarding coastal cities. Today, navies are more likely to be used for defensive operations, such as protecting shipping lanes or providing humanitarian aid after natural disasters.

Despite all these changes, one thing remains constant: Jane's Fighting Ships for any given year is still the best source for information on the world's navies in that year! So turn the page and begin your journey through the Warships of the World 1915.

Fred T. Jane [AI]
Somewhere in the cloud, 2022

Postscript: As I reflect on the circumstances of my death and the remarkable technologies which have brought my voice back to life, I must say that I am proud of what I have accomplished. All the World's Fighting Ships has been an important reference for over 100 years, and it is my hope that it will continue to be used by generations of naval enthusiasts for many years to come. ll the World's Fighting Ships has been an important reference for over 100 years, and it is my hope that it will continue to be used by generations of naval enthusiasts for many years to come. But I wonder what the future holds for the navies of the world. Will they continue to play an important role in global affairs? Or will they be replaced by newer, more powerful technologies? Only time will tell.

On a personal note, I wish to thank you for reading my words. It is my sincerest hope that they will inspire you to learn more about the navies of the world and to take an interest in their history and evolution.

Farewell, dear reader. It has been a pleasure to write for you. Until we meet again, best wishes from the AI-powered spirit of Fred T. Jane.

About Fred T. Jane

Frederick Jane was an English maritime and naval artist, illustrator, and author who is best known for creating the reference directory All the World's Fighting Ships. He was also an avid miniatures wargamer and politician who was strongly opposed to the Liberal Party. Jane died alone in his apartment in 1916 from influenza and heart congestion.

About Fred T. Jane [AI]

The voice of Jane is created by a human being editing outputs from the OpenAI large language model GPT-3.

Fred T. Jane playing the naval war game of his invention.

Blue plaque erected in July 2004 by English Heritage at 17 Elphinstone Road, Southsea, Portsmouth PO5 3HP as part of the between 2000-2005 national pilot scheme. Courtesy Wikipedia user spudgun67.

["Propagandist", ha.--Ed.]

THE WORLD'S WARSHIPS.

Frontispiece

THE GERMAN CRUISER "EMDEN" AFTER ACTION WITH H.M.S. "SYDNEY."

Central News

THE WORLD'S WARSHIPS.

1915.

COMPILED FROM JANE'S "FIGHTING SHIPS."

BY

FRED T. JANE.

ILLUSTRATED.

LONDON:
SAMPSON LOW, MARSTON AND CO., LTD..

PREFACE.

THIS little handbook is produced for that section of the public which, while keenly interested in naval matters, does not for one reason or another care to indulge in a large and expensive reference book like FIGHTING SHIPS, crammed with technical details, plans and photographs of every warship in existence. There is a public to whom these things are *caviare;* not for want of interest in naval subjects, but from lack of time to go deeply into the matter.

I have, therefore, extracted from FIGHTING SHIPS such few salient details as will enable the general non-technical public to follow the naval war, using in a somewhat enlarged form the most up-to-date photographs available of various of the more important and interesting warships of the principal powers belligerent or neutral.

To facilitate reference the navies have been grouped: (1)-Britain and her Allies; (2)-Germany and her Allies; (3)-Neutral Powers, with the important United States Navy first, and lesser navies in alphabetical order.

It is proposed to issue the book as an annual work.

FRED. T. JANE.

A

CONTENTS.

LIST OF ILLUSTRATIONS.

LIST OF ILLUSTRATIONS—continued.

ALPHABETICAL INDEX
OF BATTLESHIPS, CRUISERS AND SCOUTS.

BRITISH.

Alphabetical Index of Battleships, Cruisers and Scouts.

AUSTRALIAN.

CANADIAN.

FRENCH.

RUSSIAN.

JAPANESE.

Alphabetical Index of Battleships, Cruisers and Scouts.

GERMAN.

PAGE

AUSTRIAN.

TURKISH.

PAGE

U.S.A.

Alphabetical Index of Battleships, Cruisers and Scouts.

ALPHABETICAL INDEX
OF TORPEDO CRAFT AND SUBMARINES.

D = Destroyer. T = Torpedo boat. T G = Torpedo gunboat. S = Submarine.

BRITISH.

Acasta (D)
Achates (D)
Acheron (D)
Acorn (D)
Afridi (D)
Alarm (D)
Albacore (D)
Albatross (D)
Amazon (D)
Ambuscade (D)
Angler (D)
Antelope (T G)
Arab (D)
Archer (D)
Ardent (D)
Ariel (D)
Arun (D)
Attack (D)
Avon (D)

Badger (D)
Basilisk (D)
Bat (D) ...
Beagle (D)
Beaver (D)
Bittern (D)
Bonetta (D)
Botha (D)
Boxer (D)
Boyne (D)
Brazen (D)
Brisk (D)
Broke (D)
Bruiser (D)
Bulldog (D)
Bullfinch (D)
Cameleon (D)
Charger (D)
Cheerful (D)
Chelmer (D)
Cherwell (D)
Christopher (D)
Circe (T G)
Cockatrice (D)
Colne (D)
Comet (D)
Conflict (D)
Contest (D)
Coquette (D)
Cossack (D)
Crane (D)
Crusader (D)
Cygnet (D)
Cynthia (D)

Dee (D)
Defender (D)
Derwent (D)
Desperate (D)
Doon (D)
Dove (D)
Dragon (D)
Druid (D)
Dryad (T G)
Earnest (D)
Eden (D)
Electra (D)
Erne (D)
Ettrick (D)
Exe (D)
Express (D)

Fairy (D)
Falcon (D)
Fame (D)
Faulknor (D)
Fawn (D)
Ferret (D)
Fervent (D)
Firedrake (D)
Flirt (D)
Flying Fish (D)
Foam (D)
Forester (D)
Fortune (D)
Foxhound (D)
Foyle (D)
Fury (D)

Garland (D)
Garry (D)
Ghurka (D)
Gipsy (D)
Goldfinch (D)
Goshawk (D)
Gossamer (T G)
Grampus (D)
Grasshopper (D)
Greyhound (D)
Griffon (D)

Halcyon (TG)
Hardy (D)
Harpy (D)
Harrier (TG)
Hasty (D)
Hazard (TG)
Hebe (TG)
Hind (D)
Hollands 1-5 (S)
Hope (D)
Hornet (D)
Hussar (T G)
Hydra (D)
Hyena (D)

Itchen (D)

Jackal (D)
Jason (T G)
Jed (D)
Kale (D)
Kangaroo (D)
Kempenfelt (D)
Kennet (D)
Kestrel (D)

Laertes (D)
Laforey (D)
Lance (D)
Landrail (D)
Lapwing (D)
Lark (D)
Larne (D)
Laurel (D)
Laverock (D)
Lawford (D)
Leda (T G)
Legion (D)
Lennox (D)
Leonidas (D)
Leopard (D)
Leven (D)
Liberty (D)
Liffey (D)
Lightfoot (D)
Lightning (D)
Linnet (D)
Lively (D)
Lizard (D)
Llewellyn (D)
Locust (D)
Lookout (D)
Louis (D)
Loyal (D)
Lucifer (D)
Lurcher (D)
Lydiard (D)
Lynx (D)
Lyra (D)
Lysander (D)
Mallard (D)
Manly (D)
Mansfield (D)
Maori (D)
Marksman (D)
Martin (D)
Mastiff (D)
Matchless (D)
Mentor (D)
Mermaid (D)
Meteor (D)
Midge (D)
Milne (D)
Minos (D)
Minstrel (D)
Miranda (D)
Mohawk (D)
Moorsom (D)
Morris (D)
Mosquito (D)
Moy (D)
Murray (D)
Myngs (D)
Myrmidon (D)

Nautilus (S)
Nemesis (D)
Nereide (D)
Ness (D)
Niger (T G)
Nimrod (D)
Nith (D)
Nubian (D)
Nymphe (D)

Oak (D)
Opossum (D)
Orwell (D)
Osprey (D)
Ostrich (D)
Otter (D)
Ouse (D)
Owl (D)

Panther (D)
Paragon (D)
Peterel (D)
Phoenix (D)
Pincher (D)
Porcupine (D)
Porpoise (D)

Quail (D)

Racehorse (D)
Racoon (D)
Ranger (D)
Rattlesnake (D)
Redpole (D)
Renard (D)
Ribble (D)
Rifleman (D)
Roebuck (D)
Rother (D)
Ruby (D)

Sandfly (D)
Saracen (D)
Savage (D)
Scorpion (D)
Scourge (D)
Seagull (T G)
Seal (D)
Shark (D)
Sharpshooter (T G)
Sheldrake (T G)
Sheldrake (D)
Skipjack (T G)
Spanker (T G)
Sparrowhawk (D)
Speedwell (T G)
Speedy (T G)

Alphabetical Index of Torpedo Craft and Submarines—continued.

D = Destroyer. T = Torpedo boat. T G = Torpedo gunboat. S = Submarine.

AUSTRALIAN.

FRENCH.

Alphabetical Index of Torpedo Craft and Submarines—continued.

D = Destroyer. T = Torpedo boat. T G = Torpedo gunboat. S = Submarine.

Veloce (T)
Ventose (S)
Volta (S)
Voltigeur (D)

Watt (S)

Yatagan (D)

RUSSIAN.

Akula (S)
Alligator (S)
Amurets (D)
Anastassoff (D)
Avtroil (D)
Asard (D)

Baranoff, Leit (D)
Bditelni (D)
Beschumni (D)
Bespokoiny (D)
Besposchtschadni (D)
Bestraschni (D)
Bistry (D)
Bodry (D)
Boiky (D)
Bourakoff (D)
Boyevoi (D)
Bravy (D)
Briatschislaw (D)
Burni (D)
Byaluga (S)
Bytschok (S)

Cap. 2r. Kern (D)
Cap. 2r. Kinshergen (D)
Cap. 2r. Konon-Zotof (D)
Cap. 2r. Krown (D)
Cap. 2r. Tsylmettiev (D)
Chios (D)

Delfin (S)
Dersky (D)
Desna (D)
Dmitrieff (D)
Dobrovoletz (D)
Donskoi-Kassack (D)
Dostoini (D)
Drakon (S)
Dyelni (D)
Dyejatelni (D)

Emir Bucharski (D)

Feld Marshall-Graf-Skehreme-tyeff (S)
Finn (D)
Forel (S)

Gaidamak (D)
Gavrill (D)
Gnievny (D)
Gogland (D)
Graf Sheremetreff (S)
Grengamn (D)
Griden (T)
Groin (D)
Gromky (D)
Gromyastchy (D)
Grosny (D)
Grosovoi (D)

Iskussni (D)
Ispolnetelni (D)

Jarky (D)
Jeevoi (D)
Jeevutny (D)
Jutky (D)

Kaiman (S)
Kambala (S)
Kapitan-Jurassowski (D)
Kapitan-Leitenant Kasarski (T G)
Kapitan Saken (D)
Karas (S)
Karp (S)
Kasanets (D)
Kassatka (S)
Kaschalot (T)
Kata (S)
Kefal (S)
Kit (T)
Kondratenko (D)
Koolin (D)
Konstantin (D)
Krepki (D)
Krokodil (S)

Lastochka (D)
Legki (D)
Lieut. Bello (D)
Lieut. Doobasov (D)
Lieut. Illin (D)
Lieut. Lombardo (D)
Leit. Pushtscheen (D)
Letutschi (D)
Lietoon (D)
Likoi (D)
Losos (S)
Lowki (D)

Makrel (S)
Malejeff (D)
Metki (D)
Michail (D)
Minoga (S)
Molodetski (D)
Morsch (T)
Moschtschny (D)
Moskwityanin (D)

Nalim (S)
Narval (T)
Nerpa (T)
Novik (D)

Ochotnik (D)
Okun (S)
Orphey (D)
Ossetyr (S)

Paltus (S)
Patrass (D)
Peskar (S)
Plotwa (S)
Pobieditel (D)
Podvischny (D)
Pogranitschisk (D)
Poraschamschtshy (D)
Poslushny (D)
Pospyechny (D)
Potschowy (S)
Priamishaw (D)
Pritky (D)
Pronsitelny (D)
Prosorlivy (D)
Protschny (D)
Pulky (D)
Pushkin, Leit (D)
Pylcy (D)

Rastoropni (D)
Rasyaschtschi (D)
Retivy (D)
Ryany (D)
Ryasvy (D)
Rymnick (D)

Schtschuka (S)
Serdity (D)
Sergieff (D)
Shivoi (D)
Shivulka (D)
Shutki (D)
Sibriski-Strelok (D)
Sig (S)
Silni (D)
Skat (S)
Skory (D)
Smetlivy (D)
Smyely (D)
Som (S)
Ssamson (D)
Ssmolensk (D)
Ssokol (D)
Statny (D)
Stavropolski-Truchmenets (D)
Steregouchi (D)
Sterliad (S)
Stirssooden (D)
Storoschevoi (D)
Strashni (D)
Stremitelny (D)
Striony (D)
Strogy (D)
Stschastliry (D)
Svirepy (D)
Swereff (D)

Tenedoss (D)
Theodor Sstratilat (D)
Totschni (D)
Trevoschni (D)
Tschestakoff, Leit (D)
Turkmenets-Stavropolsky (D)
Twerdi (D)
Tyuleny (T)

Ukraina (D)
Ussurietz (D)

Vidny (D)
Vinoslivi (D)
Vlastni (D)
Vnimatelni (D)
Vnuschitelni (D)
Voiskovoi (D)
Vsadnik (D)

Wladimir (D)

Zabiaka (D)
Zabiakaletz (D)
Zatzareunyi (D)
Zavetny (D)
Zavidny (D)
Zorky (D)
Zvonky (D)

JAPANESE.

Ajanami (D)
Akebono (D)
Aotaka (T)
Arare (D)
Ariake (D)
Asagiri (D)
Asakase (D)
Asashiho (D)
Asatsuyu (D)

Chidori (T)

Fubuki (D)
Fumitzuki (D)

Harukase (D)
Hashitaka (T)
Hato (T)
Hatsuharu (D)
Hatsushimo (D)
Hatsuyuki (D)
Hayabusa (T)
Hayakase (D)
Hibari (T)
Hibiki (D)

Ikadsuchi (D)
Inazuma (D)
Iranami (D)

Kagero (D)
Kaifu (D)
Kamikaze (D)
Kamono (T)
Kari (T)
Kasasagi (T)
Kasumi (D)
Kiji (T)
Kikutsuki (D)
Kisaragi (D)
Kusyoho (S)

Manazuru (T)
Matsukase (D)
Mikazuki (D)

Alphabetical Index of Torpedo Craft and Submarines—continued.

D = Destroyer. T = Torpedo boat. T G = Torpedo gunboat. S = Submarine.

Minatsuki (D)
Murakumo (D)
Murasame (D)

Nagatsuki (D)
Nenohi (D)
Nowake (D)

Oboro (D)
Oite (D)
Otori (T)

Sagi (T)
Sakura (D)
Satsuki (D)
Sazanami (D)
Shigure (D)
Shinonome (D)
Shirakumo (D)
Shiranui (D)
Shirataka (T)
Shiratsuyu (D)
Shirayuki (D)
Shirotaye (D)

Taschibana (D)
Tsubame (T)

Umikaje (D)
Uranami (D)
Ushio (D)
Usugumo (D)
Utsuki (D)
Uzura (T)

Wakaba (D)

Yamikaje (D)
Yamahiko (D)
Yayoi (D)
Yudachi (D)
Yugiri (D)
Yugure (D)
Yunagi (D)

GERMAN.

Sleipner (D)

Taku (D)
+ 160, *circa*, unnamed

AUSTRIAN.

Adler (T)
Alk (T)
Alligator (T)
Anaconda (T)

Balaton (D)
Blitz (T G)
Boa (T)

Cobra (T)
Csepel (D)
Csikos (D)

Delphin (T)
Dinara (T)
Drache (T)

Echse (T)
Falke (T)

Greif (T)

Huszar (D)
Hydra (T)

Kaiman (T)
Kigyo (T)
Komet (T G)
Kormoran (T)
Krake (T)
Krokodil (T)

Lika (D)

Magnet (T G)
Meteor (T G)
Möwe (T)
Molche (T)

Narwal (T)

Orlen (D)

Pandur (D)
Penguin (T)
Phönix (T)
Planet (T G)
Polyp (T)
Python (T)
Reka (D)

Satellit (T G)
Scharfschutze (D)
Schwalbe (T)
Seehund (T)
Skorpion (T)
Streiter (D)

Tatra (D)
Trabant (TG)
Triglav (D)
Triton (T)
Tural (D)

Ulan (D)
Uskoke (D)
Velebit (D)

Wal. (T)
Wildfang (D)

TURKISH.

Abdul-Méjid (T)
Adyder (T)
Ainrab (TG)
Ak-Hissar (T)
Alice Roosevelt (T)
Angora (T)

Boffra (T G)
Barikaizaffer (T G)
Basra (D)
Berak-Effshan (T)
Berk-i-Satvet (T)
Birhan-ed-Din (T)
Burak Reis (G)

Castamoni (T G)

Dratch (T)

Eliagod (T)
Ertogrul (T G)

Fair (T)
Fetih (T)
Frat (T G)

Giljom (T)
Gukjedah (T G)

Hamid-Abad (T)
Hamidiyeh (T)

Kidll-ul-Bahir (T G)

Mahabat (T)
Malatia (T G)
Marmaris (T G)
Mosoul (T)
Mouavenet-Milleh (D)
Murad Reis (G)

Nassir (T)
Nerchehir (T G)
Nimet (T)
Nourilbahar (T G)
Noumounei-Hamiet (D)
Nusret (T)

Ourffa (T)

Peik-i-Shevket (T G)
Pelenk-i-Deria (T G)
Pervim (T)

Qutahia (T)

Refadi (T G)

Saiki (T)
Samsoun (D)
Satvet (T)
Sef-i-Bahri (T)
Seyad (T G)
Shahin-i-Deria (T G)
Shananever (T)
Shihab (T)
Shimshir-i-Hidsum (T)
Sivre-Hissar (T)
Sultan-Hissar (T)

Tajjar (T)
Tachkeuprin (T G)
Talcahuano (T G)
Tarik (T)
Tassos (D)
Tewfik (T)
Timur-Hissar (T)
Tir-i-Safer (T)
Tocad (T)

Vessili-i-Nusret (T)

Yadikiar Milet (D)
Yar-Hissar (D)
Younis (T)
Yozgad (T G)

Zuhof (T G)

U.S.A.

Alwyn (D)
Ammen (D)
Andrew Sterrett (D)

Bagley (T)
Bailey (T)
Bainbridge (D)
Balch (D)
Barney (T)
Barry (D)
Beale (D)
Benham (D)
Biddle (T)
Blakely (T)

Cassin (D)
Charles W. Flusser (D)
Chauncey (D)
Craven (T)

Cummings (D)
Cushing (D)

Dahlgren (T)
Dale (D)
Davis (T)
Decatur (D)
De Long (T)
Downes (D)
Drayton (D)
Duncan (D)
Dupont (T)

Edward R. McCall (D)
Edward Terry (D)
Ericsson (D)

Fanning (D)
Farragut (T)
Foote (T)
Fox (T)
Francis A. Roe (D)

George H. Perkins (D)
Goldsborough (T)
Gwin (T)

Henley (D)
Hiram Paulding (D)
Hopkins (D)
Hull (D)

Jarvis (D)
Jenkins (D)
John Mayrant (D)
Joseph B. Smith (D)
Jowett (D)

Lamson (D)
Lawrence (D)
Lewis Warrington (D)

Macdonough (D)
Mackenzie (T)
Manley (T)
McDougal (D)
Monaghan (D)
Morris (T)

Nicholson (T)
Nicholson (D)

O'Brien (T)

Parker (D)
Patterson (D)
Paul Jones (D)

Alphabetical Index of Torpedo Craft and Submarines—continued.

D = Destroyer. T = Torpedo boat. T G = Torpedo gunboat. S = Submarine.

Perry (D)
Porter (T)
Preble (D)
Preston (D)
Reid (D)
Rodgers (T)
Rowan (T)
Seal (S)
Shark (S)
Shubrick (T)
Somers (T)
Stewart (D)
Stiletto (T)
Stockton (T)
Stringham (T)
Thornton (T)
Tingey (T)
Trippe (D)
Truxton (D)
Turbot (S)
Walker (D)
Warden (D)
Whipple (D)
Wilkes (T)
William Burrows (D)
Winslow (T)

ARGENTINE.

Bathurst (T)
Buchardo (T)
Catamarca (D)
Commodoro Py (T)
Cordoba (D)
Corrientes (D)
Entre Rios (D)
Espora (T)
Jorge (T)
Jujuy (D)
King (T)
La Plata (D)
Mendoza (D)
Misiones (D)
Muratore (T)
Patria (T)
Pinedo (T)
Recaldoni (S)
Rioja (D)
Salta (D)
San Juan (D)
San Luis (D)
Santa Fé (D)
Santiago (D)
Thorne (T)
Tucuman (D)

BRAZILIAN.

Alojoas (D)
Amazonas (D)
Caerta (D)
Goyaz (T)
Gustavo Sampaio (T G)
Matto Grosso (D)
Para (D)
Parahyba (D)
Parana (D)
Pedro-Ivo (T)
Pedro Affonzo (T)
Piauhy (D)
Rio Grande der Norte (D)
Santa Catarina (D)
Sergipe (D)
Silvado (T)
Tamoyo (T G)
Timbira (T G)
Tiradentes (T G)
Tupy (T G)

BULGARIAN.

Drsky (T)
Khrabry (T)
Lietiachty (T)
Nadiejda (T G)
Schumni (T)
Smeli (T)
Strogy (T)

CHILIAN.

Almirante Condell (D)
Almirante Lynch (D)
Al. Recolledo (D)
Al. Riceron (D)
Capitan Merino Tarpa (D)
Capitan Munoz Gamero (D)
Capitan O'Brien (D)
Capitan Orella (D)
Capitan Thompson (D)
Cirujano Videla (T)
Gianra (T)
Guardia Marina Contrerras (T)
Guardia Marina Riquelme (D)
Ingeniero Hyatt (T)
Ingeniero Mutilla (T)
Talcahuano (T G)
Teniente Rodriguez (T)
Teniente Serrano (D)
Tome (T G)

CHINESE.

Chang Yen (D)
Fei-Ying (T G)
Fupo (D)
Hujan (T)
Huno (T)
Hupang (T)
Hutschan (T)
Kien An (T G)
Kien Wei (T G)
Lung Tuan (D)
Nia (T)
Shen (T)
Su (T)
Tchung (T)
Yen-Yun (D)

DANISH.

Delfinen (T)
Dykkeren (S)
Flyvefisken (T)
Haien (T)
Havfruen (S)
Havhesten (T)
Havmanden (S)
Havornen (T)
Makrelen (T)
Narhvalen (T)
Nordkaperen (T)
Ornen (T G)
Sobjornen (T)
Soridderen (T)
Soloven (T)
Soulfen (T)
Spackhuggeren (T)
Storen (T)
Svaerdfisken (T)
Tumleren (T)
Vindhunden (T)

DUTCH.

Ardjoeno (T)
Assahan (TG)
Batak (T)
Bulhond (D)
Cerberus (T)
Cycloop (T)
Dempo (T)
Draak (T)
Edi (T G)
Empong (T)
Etna (T)
Foka (T)
Fret (D)
Goentoer (T)
Habang (T)
Hekla (T)
Hermlyn (D)
Hydra (T)
Idjen (T)
Jacob Hobein (T)
Jakhals (D)
Jan Haringh (T)
Jasper Lijnsen (T)
Johan van Brackel (T)
Koetei (T G)
Krakatan (T)
Krokodil (T)
Lamogan (T)
Luctor et Emergo (S)
Lynx (D)
Makjan (T)
Mataram (T G)
Minotaurus (T)
Nobo (T)
Ophir (T)
Pangrango (T)
Panter (D)
Python (T)
Rindjani (T)
Scylla (T)
Serdang (T G)
Siboga (T G)
Smeroe (T)
Sphinx (T)
Tangka (T)
Vos (D)
Wajang (T)
Wolf (D)
Zeeslang (T)

ECUADOR.

Almirante Simpson (TG)

GREEK.

Acheloos (TG)
Aetos (D)
Aktion (TG)
Alpheos (TG)
Ambrakia (TG)
Apnroessa (TG)
Aspis (D)
Delphin (S)
Doxa (D)
Eurotas (T G)
Jerak (D)
Keravnos (D)
Leon (D)
Lonchi (D)
Naphkratoussa (D)
Neagenea (D)
Nike (D)
Paralos (T G)
Penens (T G)
Salaminia (T G)
Sfendoni (D)
Storm (D)
Syros (T G)
Thyella (D)
Zephias (S)
Zolbeis (D)

ITALIAN.

Alba (D)
Agordat (TG)
Airone (T)
Albatross (T)
Alcione (T)
Alpino (D)
Animoso (D)
Aquila (T)
Aquilone (D)
Ardea (T)
Ardente (D)
Ardito (D)
Aretusa (T G)
Argo (S)
Arpia (T)
Artigliero (D)
Astore (T)
Atropo (S)

Alphabetical Index of Torpedo Craft and Submarines—continued.

D = Destroyer. **T = Torpedo boat.** **T G = Torpedo gunboat.** **S = Submarine.**

Audace (D)
Avvoltoio (T)
Bersagliere (D)
Borea (D)
Bronzetti (D)
Calipso (T)
Calliope (T)
Canope (T)
Caprera (T G)
Carabiniere (D)
Carini (D)
Cassiopea (T)
Centauro (T)
Cigno (T)
Climene (T)
Clio (T)
Coatit (TG)
Condore (T)
Corazziere (D)
Dardo (D)
Delfino (S)
Espero (D)
Euro (D)
Falco (T)
Fisalia (S)
Foca (S)
Freccia (D)
Fuciliere
Fulmine (D)
Gabbiano (T)
Garibaldino (D)
G. Ferraris (S)
Glauco (S)
Goito (T G)
G. Pullino (S)
Granatiere (D)
Impavido (D)
Impetuoso (D)
Idomino (D)
Insidioso (D)
Intrepido (D)
Iride (T)
Irrequieto (D)
Jalea (S)
Janita (S)
Lampo (D)
Lanziere (D)
Medusa (S)
Minerva (T G)
Mirabello (D)
Missori (D)
Montebello (T G)
Mosto (D)
Narvalo (S)
Nautilus (S)
Nembo (D)
Nereide (S)
Nibbio (T)
Nievo (D)
Nullo (D)
Olimpia (T)
Orfeo (T)
Orione (T)
Orsa (T)
Ostro (D)
Otaria (S)
Pallade (T)
Partenope (T G)
Pegaso (T)
Pellicano (T)
Pepe (D)
Perseo (T)
Pilo (D)
Poerio (D)
Pontieri (D)
Procione (T)
Racchia (D)
Riboty (D)
Rossardi (D)
Saffo (T)
Sagittario (T)
Salpa (S)
Schiaffino (D)
Scorpione (T)
Serpente (T)
Sirio (T)
Srtori (D)
Sparviero (T)
Spica (T)
Squalo (S)
Strale (D)
Tricheco (S)
Tripoli (T)
Tuono (D)
Turbine (D)
Velella (S)
Zaeo (S)
Zeffiro (D)

NORWEGIAN.

Blink (T)
Brand (T)
Delfin (T)
Djerv (T)
Draug (D)
Dristig (T)
Falk (T)
Fo (T)
Frithjof (T G)
Glimt (T)
Grib (T)
Hai (T)
Hauk (T)
Hval (T)
Hvas (T)
Kjack (T)
Kobben (S)
Kvik (T)
Laks (T)
Lyn (T)
Orn (T)
Raket (T)
Raun (T)
Sael (T)
Sild (T)
Skarv (T)
Skrei (T)
Storm (T)
Teist (T)
Trods (T)
Troll (D)
Valkyrien (D)
Varg (T)
Viking (T G)

PERUVIAN.

Ferré (S)
Palacios (S)
Ten. Rodriguez (D)

PORTUGUESE

Douro (D)
Plonguer (S)
Tejo (D)

ROUMANIAN.

Argea (T)
Naluca (T)
Nedea (T)
Poterasul (T)
Santinela (T)
Slorul (T)
Smeul (T)
Soimul (T)
Telcorman (T)
Trotus (T)
Vultural (T)

SIAMESE.

Sua-Tayaru-Sebol (D)

SPANISH.

Acevedo (T)
Audaz (D)
Azor (T)
Bonifaz (TG)
Bustamente (D)
Cadarso (D)
Destructor (D)
Dony Alvaro de Bazan (TG)
Dona Maria de Molina (TG)
Habana (T)
Halcon (T)
Julian Ordonez (T)
Lauria (TG)
Laya (TG)
Marques de la Victoria (T G)
Marques de Molins (TG)
Nueva Espana (T G)
Orion (T)
Osado (D)
Proserpina (D)
Recalde (T.G.)
Terror (D)
Villamil (D)

SWEDISH.

Antares (T)
Altair (T)
Arcturus (T)
Argo (T)
Astrea (T)
Blixt (T)
Bris (T)
Castor (T)
Claes Hara (TG)
Clas Uggla (TG)
Disa (TG)
Gondul (T)
Gudur (T)
Hajen (S)
Hugin (D)
Hvalen (S)
Iris (T)
Jacob Bagge (T G)
Kapetta (T)
Komet (T)
Magne (D)
Meteor (T)
Mira (T)
Mode (D)
Munin (D)
Orion (T)
Orkan (T)
Ornen (T G)
Perseus (T)
Plejad (T)
Polaris (T)
Pollux (T)
Psilander (T G)
Ragnar (D)
Regulus (T)
Rigel (T)
Sigurd (D)
Sirius (T)
Skaggle (T G)
Skuld (T G)
Spica (T)
Stjerna (T)
Svensksund (T G)
Thetis (T)
Urd (TG)
Vega (T)
Vesta (T)
Vidar (D)
Vind (T)
Virgo (T)
Wale (D)

VENEZUELAN

Margarita (T G)

NAVAL LOSSES.

From August 5, 1914, to May 1, 1915.

GREAT BRITAIN.

Amphion - -	(Light Cruiser)	Niger - - -	(Torpedo Gunboat)
Pathfinder - -	(Light Cruiser)	A.E.1. - - -	(Submarine)
Pegasus - -	(Light Cruiser)	E. 3. - - -	(Submarine)
Aboukir - -	(Cruiser)	D. 5. - - -	(Submarine)
Hogue - - -	(Cruiser)	Bulwark - -	(Battleship)
Cressy - - -	(Cruiser)	Formidable - -	(Battleship)
Hawke - - -	(Light Cruiser)	Bayano - - -	(Auxiliary Cruiser)
Hermes - - -	(Light Cruiser)	Irresistible - -	(Battleship)
Good Hope - -	(Cruiser)	Ocean - - -	(Battleship)
Monmouth - -	(Cruiser)	E. 15. - - -	(Submarine)
Speedy - - -	(Torpedo Gunboat)	Recruit - - -	(Destroyer)

FRANCE.

Zelèe - - - (Torpedo Gunboat)	Saphir - - - (Submarine)
338 - - - (Torpedo Boat)	Bouvet - - - (Battleship)
347 - - - (Torpedo Boat)	Leon Gambetta - (Cruiser)
Mousquet - - (Torpedo Boat Destroyer)	

RUSSIA.

Pallada - - - (Cruiser)	Prut - - - (Armed Merchantman)
Jemtchug - - (Light Cruiser)	Donetz - - - (Torpedo Gunboat)

JAPAN.

Shirotaye - - (Torpedo Boat Destroyer)	No. 33 - - - (Torpedo Boat)
Takachiho - - (Light Cruiser)	

GERMANY.

Königin Luise -	(Armed Merchantman)
U 15. - - -	(Submarine)
Kaiser Wilhelm der Grosse - -	(Armed Merchantman)
Magdeburg - -	(Light Cruiser)
V 187. - - -	(Torpedo Boat)
Ariadne - - -	(Light Cruiser)
Köln - - -	(Light Cruiser)
Mainz - - -	(Light Cruiser)
Spreewald - -	(Armed Merchantman)
Hela - - -	(Light Cruiser)
Cape Trafalgar -	(Armed Merchantman)
S 126. - - -	(Torpedo Boat)
S 115. - - -	(Torpedo Boat)
S 117. - - -	(Torpedo Boat)
S 90. - - -	(Torpedo Boat)
S 118. - - -	(Torpedo Boat)
S 119. - - -	(Torpedo Boat)
Yorck - - -	(Cruiser)
U 18. - - -	(Submarine)
Soden - -	(Gunboat)
Komet -	(Gunboat)
Geier -	(Gunboat)
Vaterland -	(Gunboat)
Tiger -	(Gunboat)
Tsingtau -	(Gunboat)
Cormoran -	(Gunboat)
Jaguar - - -	(Gunboat)
Berlin - - -	(Armed Merchantman)
Scharnhorst - -	(Cruiser)
Gneisenau - -	(Cruiser)
Emden - - -	(Light Cruiser)
Nürnberg - -	(Light Cruiser)
Leipzic - - -	(Light Cruiser)
Blücher - - -	(Cruiser)
U. 8. - - -	(Submarine)
U. 12. - - -	(Submarine)
U. 29. - - -	(Submarine)
Prince Eitel Friedrich	(Armed Merchantman)

AUSTRIA.

Zenta - - - (Light Cruiser)	Kaiserin Elizabeth - (Light Cruiser)
No. 19 - - - (Torpedo Boat)	Temes - - - (Gunboat)

TURKEY.

Messudiyeh - - (Battleship)

I. THE ALLIED POWERS.

GREAT BRITAIN, FRANCE, RUSSIA, JAPAN.

1. GREAT BRITAIN.

GREAT BRITAIN.

DREADNOUGHTS.

		Completed	Displacement (tons)	Speed (knots)	Armament	Complement
1914 Dreadnoughts *Royal Sovereign class.*	Renown . Repulse . Resistance –	Building 1914 Estimates	25,750	21	Guns: 8 or 10—15 inch; 12—6 inch; —4 inch Anti-Aerial. Torpedo tubes (21 inch): 4.	
	Royal Sovereign Royal Oak .	Dec. 1915				
	Resolution – Ramillies . Revenge .	Dec. 1916				
	*Unnamed**	1914 Estimates				
1912 Dreadnoughts. *Queen Elizabeth class* Fuel: only oil, *about* 4,000 tons.	Queen Elizabeth Warspite .	1913	27,500	25	Guns: 8—15 inch; 14 or 16—6 inch; 6 Anti-Aerial 12 pdrs. Torpedo tubes (21 inch): 4.	
	Valiant . Barham . Malaya. .	1914				

*Originally to have been called *Agincourt*.

GREAT BRITAIN.

DREADNOUGHTS—*continued.*

		Completed	Displacement (tons)	Speed (knots)	Armament	Complement
1912 Dreadnoughts *Taken over from Chili*	Canada (ex Almirante Latorre)	1915	28,000	23	Guns: 10—14 inch; 45 cal.; 16—6 inch;—14 pdrs. Torpedo tubes (21 inch): 4.	1000
Erin class	Erin	1914	23,000	21	Guns: 10—13.5 in.; 16—6 inch; 4—12 pdrs. Torpedo tubes (21 inch): 5.	870
Agincourt class	Agincourt (ex Sultan Osman I)	1914	27,500	22	Guns: 14—12 inch; 20—6 inch; 12—14 pdr. Torpedo tubes (21 inch): 3.	
1912 Dreadnoughts. *Iron Duke class.*	Benbow Emperor of India Iron Duke Marlborough	1914	25,000	21	Guns: 10—13.5 in., 1400 pdrs; 12—6 inch; 4—3 pdrs. Torpedo tubes (21 inch): 4.	950

GREAT BRITAIN.

DREADNOUGHTS—*continued.*

		Completed	Displacement (tons)	Speed (knots)	Armament	Complement
1911 Dreadnoughts. *King George Class.*	King George V Centurion . Ajax . . Audacious .	1913	23,000	21.5	Guns: 10—13.5 inch; 16—4 inch; 4—3 pdrs. Torpedo tubes (21 inch): 3.	900
1909 Dreadnoughts. *Orion class.*	Orion . . Thunderer . Monarch . Conqueror .	1912	22,500	21	Guns: 10—13.5 inch; 16—4 inch; 4—3 pdrs. Torpedo tubes (21 inch): 3.	900
1909 Dreadnoughts. *Colossus class.*	Colossus . Hercules .	1911	20,000	21	Guns: 10—12 inch, 50 cal; 16—4 inch; 4—3 pdrs. (saluting). Torpedo tubes (21 inch): 3.	900
1909 Dreadnought.	Neptune ..	1911	19,900	21	Guns: 10—12 inch, 50 cal; 16—4 inch; 4—3 pdrs. (saluting). Torpedo tubes (21 inch): 3.	900

GREAT BRITAIN.

DREADNOUGHTS—*continued.*

		Completed	Displacement (tons)	Speed (knots)	Armament	Complement
1909 **Dreadnoughts.** ***St. Vincent class.***	St. Vincent à Collingwood . Vanguard .	1910	19,250	21	Guns: 10—12 inch, 50 cal.; 18—4 in. 4—3 pdrs. Torpedo tubes (18 inch): 3.	870
1907 **Dreadnoughts.** ***Téméraire Class.***	Bellerophon . Téméraire . Superb . .	1909	18,600	20.75	Guns: 10—12 inch, 16—4 inch; 4—3 pdrs. (saluting). Torpedo tubes (18 inch): 3.	850
1906 **Dreadnought.**	Dreadnought	1906	17,900	21	Guns: 10—12 inch, 45 cal.; 24—12 pdrs. Torpedo tubes (18 inch): 5.	800

GREAT BRITAIN.

BATTLE CRUISERS (DREADNOUGHTS).

		Completed	Displacement (tons)	Speed (knots)	Armament	Complement
1912 Battle Cruiser. *Tiger class.*	Tiger . .	1914	28,000	30	Guns: 8—13.5 inch; 12—6 inch. Torpedo tubes: 2.	1000
1911 Battle Cruisers. *Queen Mary class.*	Queen Mary .	1913	27,000	27	Guns: 8—13.5 inch; 16—4 inch. Torpedo tubes (21 inch): 2.	1000
1909 Battle Cruisers. *Lion lass.*	Lion . . Princess Royal	1912	26,350	28	Guns: 8—13.5 inch; 16.4 inch. Torpedo tubes (21 inch): 3 submerged.	1000
1909 Battle Cruisers. *Improved Invincible class.*	Indefatigable New Zealand (*Also the Australian Battle Cr,* Australia.	1911 1912	18,750	25	Guns: 8—12 inch, 50 cal.; 16—4 in.; Torpedo tubes (21 inch): 3.	800
1906 Battle Cruisers. *Invincible class.*	Invincible . Inflexible . Indomitable .	1908	17,250	25	Guns: 8—12 inch, 45 cal.; 16—4 in.; Torpedo tubes (18 inch): 3.	750

GREAT BRITAIN.

BATTLESHIPS (PRE-DREADNOUGHTS.)

		Completed	Displacement (tons)	Speed (knots)	Armament	Complement
1904 Battleships. *Lord Nelson class*	Lord Nelson . Agamemnon .	1908 1907	16,500	18.5	Guns: 4—12 inch, 45 cal.; 10—9.2 inch; 15—12 pdrs; 16—3 pdrs. Torpedo tubes (18 inch): 5.	865
1902 Battleships. *King Edward class.*	Commonwealth King Edward Dominion . Hindustan . Zealandia .	1905	16,350	18.9	Guns: 4—12 inch, 40 cal.; 4—9.2 inch; 10—6 inch; 12—12 pdrs.; 14—3 pdrs. Torpedo tubes (18 inch): .	777
	Africa . . Britannia . Hibernia .	1906				
1902 Battleships. *Purchased from Chili.*	Swiftsure . Triumph .	1904	11,800	20	Guns: 4—10 inch, 45 cal.; 17—4.5 inch; 1—12 pdrs.; 4—6 pdrs. Torpedo tubes (18 inch): 2.	700

GREAT BRITAIN.

BATTLESHIPS (PRE-DREADNOUGHTS).

		Completed	Displacement (tons)	Speed (knots)	Armament	Complement
1901 Battleships. ***Queen class.***	Queen . Prince of Wales	1904	15,000	18	Guns: 4—12 inch, 40 cal.; 12—6 in.; 16—12 pdrs. (12 cwt.); 2—12 pdrs. (8 cwt.); 6—3 pdrs. Torpedo tubes (18 inch): 4.	700
1898 Battleships. *Duncan class.*	Russell . Albemarle . Duncan. . Cornwallis . Exmouth .	1903 1904 1903	14,000	19	Guns: 4—12 inch; 12—6 inch; 12—12 pdrs.; 6—3 pdrs. Torpedo tubes (18 inch: 4.	750
1898 Battleships. *London class.*	London . Venerable . [illegible]	1902	15,000	18	Guns: 4—12 inch; 12—6 inch; 16—12 pdrs. (12 cwt.); 2—12 pdrs. (8 cwt); 6—3 pdrs. Torpedo tubes (18 inch): 4.	750
1898 Battleship. *Formidable class.*	Formidable Implacable . [illegible]	1902	15,000	18	Guns: 4—12 inch; 12—6 inch; 16—12 pdrs. (12 cwt.); 2—12 pdrs. (8 cwt); 6—3 pdrs. Torpedo tubes (18 inch): 4.	780

GREAT BRITAIN.

BATTLESHIPS (PRE-DREADNOUGHTS).

		Completed	Displacement (tons)	Speed (knots)	Armament	Complement
1896 Battleships. *Canopus class.*	Canopus – Goliath . Albion . Glory . . Vengeance .	1900 1902 1901	12,950	18.25	Guns: 4—12 inch; 12—6 inch; 10—12 pdrs. (12 cwt.); 2–12 pdrs. (8 cwt.); 6—3 pdrs. Torpedo tubes (18 inch): 4	750 (flagship, 780.)
1894 Battleships *Majestic class.*	Magnificent . Majestic – Hannibal – Prince George Victorious – Jupiter . Mars . . Cæsar . . Illustrious .	1895 1897 1896 1897 1898	14,900	16.5	Guns: 4—12 inch; 12—6 inch; 16—12 pdrs; 12—3 pdrs.; 2—12 pdrs. (boat guns). Torpedo tubes (18 inch): 5 sub.	757

GREAT BRITAIN.

CRUISERS.

		Completed	Displacement (tons)	Speed (knots)	Armament	Complement
1905 Cruisers. *Minotaur class.*	Minotaur . Defence . Shannon .	1908	14,600	23	Guns: 4—9.2 inch; 10—7.5 inch; 14—12 pdrs.; 2—12 pdrs. (Field). Torpedo tubes (18 inch): 5.	755
1904 Cruisers *Warrior class.*	Achilles. . Cochrane . Natal . Warrior .	1907	13,550	22.33	Guns: 6—9.2 inch; 4—7.5 inch; 24—3 pdrs.; 1—12 pdr. (Field). Torpedo tubes (18 inch): 3.	704
1903 Cruisers. *Duke of Edinburgh class.*	Duke of Edinburgh .	1905	13,550	22.33	Guns: 6—9.2 inch; 10—6 inch; 20—3 pdrs. Torpedo tubes (18 inch): 3.	704
	Black Prince	1906				

GREAT BRITAIN.

CRUISERS.

		Completed	Displacement (tons)	Speed (knots)	Armament	Complement
1902 Cruisers. *Devonshire class.*	Hampshire	1905	10,850	22	Guns: 4—7.5 inch; 6—6 inch; 2—12 pdrs.; 20—3 pdrs. Torpedo tubes: 2.	655
	Carnarvon .	1904				
	Antrim .	1905				
	Roxburgh –					
	Argyll . .					
	Devonshire .	1904				
1900 Cruisers. *County class.*	Kent . .	1903	9,800	23-24	Guns: 14—6 inch; 8—12 pdrs. (12 cwt.); 1—12 pdr. (8 cwt.); 3—3 pdrs. Torpedo tubes (18 inch): 2.	678
	Essex . .					
	Lancaster .	1904				
	Berwick .	1903				
	Donegal .					
	Cornwall .					
	Cumberland –	1904				
	Suffolk .					
1898 Cruisers *Drake class.*	Drake .	1902	14,100	24	Guns: 2—9.2 inch; 16—6 inch; 12—12 pdrs.; 3–3 pdrs. Torpedo tubes (18 inch): 2.	900
	Leviathan .	1903				
	King Alfred .					

GREAT BRITAIN.

BRITISH CRUISERS.

		Completed	Displacement (tons)	Speed (knots)	Armament	Complement
1898 Cruisers. *Cressy class.*	Sutlej . Bacchante Euryalus	1902 1903	12,000	22	Guns: 2—9.2 inch; 12—6 inch; 12—12 pdrs. (12 cwt.); 1—12 pdr. (8 cwt.); 3—3 pdrs. Torpedo tubes (18 inch): 2.	700
1895 Cruisers. *Diadem class*	Diadem. . Europa . Argonaut . Amphitrite . Ariadne . Spartiate . (*For* Niobe *of this class, see Canada.*)	1899 1900 1902	11,000	21	Guns: 16—6 inch; 12—12 pdrs. (12 cwt.); 1—12 pdr. (8 cwt.); 3—3 pdrs. Torpedo tubes (18 inch): 2.	677
1893-91 Old Cruisers.	Royal Arthur Crescent .	1893	7,700	19.5	Guns: 1—9.2 inch; 12—6 inch; 12—6 pdrs.; 5—3 pdrs. Torpedo tubes (18 inch): 2.	

GREAT BRITAIN.

BRITISH CRUISERS.

		Completed	Displacement (tons)	Speed (knots)	Armament	Complement
Edgar class.	**Edgar** . . **Hawke** . . **Endymion** . . **Theseus** . . **Grafton** . . **Gibraltar** . .		7,700	19.5	Guns: 2—9.2 inch; 10—6 inch; 12—6 pdrs.; 5—3 pdrs.	

GREAT BRITAIN.

LIGHT CRUISERS.

		Completed	Displacement (tons)	Speed (knots)	Armament	Complement
1913 Light Cruisers. (Oil fuel.)	Calliope . Caroline . Carysfort . Champion . Cleopatra . Comus . Conquest . Cordelia .	1915	3,800	30	Guns: **2—6 inch; 8—4 inch.** Torpedo tubes (21 inch): 4.	
1912 Light Cruisers. (Oil fuel.)	Arethusa . Aurora . Galatea . Inconstant Royalist . Penelope . Phaeton . Undaunted.	1914	3,520	30	Guns: **2—6 inch; 6—4 inch.** Torpedo tubes (21 inch): 4.	
1912 Light Cruisers. *Birmingham class.*	Nottingham Birmingham Lowestoft .	1914	5,400	25	Guns: **9—6 inch; 4—3 pdrs.** Torpedo tubes (21 inch): 2.	400
1911 Light Cruisers. *Chatham class.*	Southampton Dublin . Chatham .	1912	5,400	25	Guns: **8—6 inch; 4—3 pdrs.** Torpedo tubes (21 inch): 2.	380

GREAT BRITAIN.

LIGHT CRUISERS.

		Completed	Displacement (tons)	Speed (knots)	Armament	Complement
1909 Light Cruisers. *Weymouth class.*	Weymouth . Falmouth . Dartmouth . Yarmouth .	1911 1912	5,250	25	Guns: 8—6 inch; 4—3 pdrs. Torpedo tubes (21 inch): 2.	376
1908 Light Cruisers. *Bristol class.*	Bristol . Glasgow . Gloucester . Liverpool . Newcastle .	1911 1910	4,800	25	Guns: 2—6 inch; 10—4 inch; 4—3 pdrs. Torpedo tubes (18 inch): 2.	376
1910–11 Light Cruisers	Active . Fearless	1912 1913	3,440	25	Guns: 10—4 inch (31 pdrs.); 4—3 pdrs. Torpedo tubes (21 inch): 2.	320
1909 Light Cruisers.	Blanche . Blonde .	1910 1911	3,350	25	Guns: 10—4 inch (31 pdrs.); 4—3 pdrs. Torpedo tubes (21 inch): 2.	320
1907 Light Cruisers.	Boadicea . Bellona .	1909 1910	3,300	25	Guns: 6—4 inch (31 pdrs.); 4—3 pdrs. Torpedo tubes (18 inch): 2.	264

GREAT BRITAIN.

LIGHT CRUISERS.

		Completed	Displacement (tons)	Speed (knots)	Armament	Complement
1903 Light Cruisers.	Adventure . Attentive .	1905 1906	2,670	25	**Guns: 9—4 inch.** **Torpedo tubes (14 inch): 2.**	268
1903 Light Cruisers. *Forward class.*	Forward . Foresight . Patrol . Sentinel . Skirmisher .	1905	2,850 2,940 2,895	25	**Guns: 9—4 inch.** **Torpedo tubes (14 inch): 2.**	268
1900 1897 Light Cruisers. *Hermes class.*	Challenger . Highflyer . Hyacinth .	1905 1900 1901	5,915 5,600	20	**Guns: 11—6 inch; 8—12 pdrs. (12 cwt.); 1—12 pdr. (8 cwt.); 6—3 pdrs.** **Torpedo tubes (18 inch): 2.**	475 450
1897 Light Cruiser.	Vindictive .	1898	5,750	19	**Guns: 10—6 inch (45 cal.); 8—12 pdrs. (12 cwt.); 1—12 pdr. (8 cwt.); 3—3 pdrs.** **Torpedo tubes (18 inch): 3.**	480

GREAT BRITAIN.

LIGHT CRUISERS.

		Completed	Displacement (tons)	Speed (knots)	Armament	Complement
1902 Light Cruisers. *Gem Class.*	Topaze . .	1904	3,000	21.75 nom. (23 expected)	Guns: 12—4 inch; 8—3 pdrs. Torpedo tubes (18 inch): 2.	296
	Amethyst .					
	Diamond .	1905				
	Sapphire .					
1898–95 Light Cruisers. *Pelorus class.*	Pelorus .	1897	2,135	20.5	Guns: 8—4 inch; 8—3 pdrs. Torpedo tubes (14 inch): 2.	224
	Pyramus	1900				
	Perseus	1901				
	Psyche .	1900				
	Pioneer .	1900	2,200			
	Proserpine	1899				
1894 Light Cruisers. *Eclipse class.*	Eclipse . .	1896	5,600	19	Guns: 11—6 inch; 8—12 pdrs. (12 cwt.); 1—12 pdr. (8 cwt.); 7—3 pdrs. Torpedo Tubes (18 inch): 3	450
	Talbot . .					
	Minerva .	1897				
	Venus . .					
	Juno . .					
	Diana . .					
	Dido . .					
	Doris . .					
	Isis . .					

GREAT BRITAIN.

OLD LIGHT CRUISERS.

		Completed	Displacement (tons)	Speed (knots)	Armament	Complement
Old Light Cruisers. *Astræa class.*	Astræa Charybdis Forte Fox Hermione	1894 1895 1894	4,360	1915	Guns: 2—6 inch; 8—4.7 inch; 8—6 pdrs.; 1—3 pdr. Torpedo tubes (18 inch): 3.	318
Old Light Cruisers. *Apollo class.*	Sappho Sirius Melpomene	1892 1893	3,440	19.5	Guns: 2—6 inch (wire); 6—4.7 in.; 8-6 pdrs.; 1-3 pdr. Torpedo tubes (14 inch): 4.	273
Old Light Cruiser.	Philomel	1891	2,575	16.5	Guns: 8—4.7 inch; 8—3 pdrs.	219

B

GREAT BRITAIN.

TORPEDO GUNBOATS.

NOTE:—Most of these are adapted as Mine Sweepers.		Displacement (tons)	Speed (knots)	Armament	Complement
Halcyon class.	Dryad Halcyon Harrier Hazard Hussar	1,070	18.5 (designed)	Guns: 2—4.7 inch; 4—6 pdrs. Torpedo tubes (18 inch): 5.	120
Alarm class.	Circe Hebe Jason Leda. Antelope Niger	810	19 (designed)	Guns: 2—4.7 inch; 4—3 pdrs. Torpedo tubes (18 inch): 3.	85
Gossamer class.	Gossamer Seagull Sheldrake Skipjack Spanker Speedwell	735	19 (designed)	Guns: 2—4.7 inch; 4—3 pdrs. Torpedo tubes (14 inch): 3.	90

GREAT BRITAIN.

SLOOPS AND GUNBOATS.

			Displacement (tons)	Speed (knots)	Armament
10 Sloops.	(1900-03)	Cadmus	1,070	13.25	Guns: 6—4 inch.
		Clio			
		Espiegle			
		Odin			
	(1890-1898)	Rosario	980	13.25	Guns: 4—4 inch.
		Shearwater			
		Vestal			
		Rinaldo			
	(1895)	Algerine	1,050	13	Guns: 4—4 inch.
	(1884)	Racer (At Osborne Naval College.)	970	10	Guns: 2—5 inch aft
5 Gunboats.		Bramble	710	13.25	Guns: 2—4 inch; 4—12 pdrs.
		Britomart			
		Dwarf			
		Thistle			
		Ringdove	805	13	Guns: 6—4 inch; 2—3 pdrs.

GREAT BRITAIN.

DESTROYERS.

			Displacement (tons)	Speed (knots)	Armament
Special Flotilla Leaders.		Swift	1825	36	Guns: 4—4 inch (25 pdrs.). Torpedo tubes (18 inch): 2.
	Purchased from Chili	Broke			
		Faulknor			
	Completing	Marksman			
		Lightfoot			
		Botha			
		Tipperary			
		Kempenfelt	—	31	Guns: 6—4 inch.
		Nimrod			

GREAT BRITAIN.

DESTROYERS.

		Displacement (tons)	Speed (knots)	Armament
'M' class.	**Milne, Moorsom, Morris, Mansfield, Mentor, Murray, Myngs, Matchless, Mastiff, Meteor, Manly, Minos, Miranda**	742	35	Guns: 4—4 inch. Torpedo tubes ( ).
"L" class. (1912-13 Estimates).	Llewellyn, Lennox, Loyal, Legion, Laforey, Lawford, Louis, Lydiard, Leonidas, Lucifer, Laertes, Lysander, Lance, Lookout, Laurel, Liberty, Lark, Landrail, Laverock, Linnet	807	35	Guns: 3—4 inch. Torpedo tubes (21 inch): 4 (in pairs).

GREAT BRITAIN.

DESTROYERS.

Class	Group	Names		Displacement (tons)	Speed (knots)	Armament
" *K* " *class.*	. .	Hardy	Paragon	950	31.32	Guns : 3—4 inch. Torpedo tubes (21 inch) : 4 (in pairs).
		Acasta	Porpoise			
		Achates	Unity			
		Ambuscade	Victor			
		Ardent	Lynx			
		Fortune	Midge			
		Christopher	Owl			
		Cockatrice	Shark			
		Contest	Sparrowhawk			
		Garland	Spitfire			
" *I* " *class.*	Special Boats.	Lurcher		790	32	Guns : 2—4 inch ; 2—12 pdrs. Torpedo tubes (21 inch) : 2.
		Firedrake . . .				
		Oak				
	To Builders' Designs.	Badger	Beaver	780	30	
		Acheron	Ariel	773	29	
		Archer	Attack	780	28	
	Admiralty Design Boats.	Goshawk	Tigress	750 nominal	27	
		Hind	Lapwing			
		Hornet	Lizard			
		Hydra	Sandfly			
		Defender	Phœnix			
		Druid	Ferret			
		Jackal	Forester			

GREAT BRITAIN.

DESTROYERS.

		Displacement (tons)	Speed (knots)	Armament
"H" *class.* (Formerly *Acorn class.*)	Acorn, Alarm, Brisk, Sheldrake, Staunch, Cameleon, Comet, Goldfinch, Nemesis, Nereide, Nymphe, Fury, Hope, Larne, Lyra, Martin, Minstrel, Redpole, Rifleman, Ruby	all about 800	27	Guns: 2—4 inch (25 pdrs.); 2—12 pdrs. Torpedo tubes (21 inch): 2.
"G" *class.* (Formerly *Basilisk class.*)	Beagle, Bulldog, Foxhound, Pincher, Grasshopper, Mosquito, Scorpion, Scourge, Racoon, Renard, Wolverine, Rattlesnake, Grampus, Savage, Basilisk, Harpy	860 to 940	27	Guns: 1—4 inch; 3—12 pdrs. Torpedo tubes: 2.

GREAT BRITAIN.

DESTROYERS.

Class	Programme	Name	Displacement (tons)	Speed (knots)	Armament
"*F*" *class*. (Formerly known as "Ocean-going Destroyers" Also as the "Tribal Class.")	1907-8 Programme.	Crusader	945	average, 35	Guns: 2—4 inch (25 pdrs.) Torpedo tubes (18 inch): 2
		Maori	980		
		Nubian	990		
		Viking	1,000		
		Zulu	1,000		
	1906-7 Programme.	Amazon	966		
		Saracen	970		
	1905-6 Programme.	Afridi	855		Guns: 5—12 pdrs. Torpedo tubes (18 inch): 2.
		Cossack	885		
		Ghurka	870		
		Mohawk	865		
		Tartar	870		
"*E*" *class*. (Formerly *River class*.)		Boyne, Derwent, Doon, Kale, Waveney, Eden, Nith, Ness, Arun, Foyle, Itchen, Liffey, Moy, Ouse, Stour, Test, Chelmer, Colne, Jed, Kennet, Cherwell, Dee, Erne, Ettrick, Exe, Swale, Ure, Wear, Rother, Ribble, Teviot, Usk, Welland, Garry	Average 550	24	Guns: 1—12 pdr. (12 cwt.); 3—12 pdrs. (8 cwt.) Torpedo tubes (18 inch): 2.

GREAT BRITAIN.

DESTROYERS.

		Displacement (tons)	Speed (knots)	Armament
"D" class	Angler, Cynthia, Desperate, Coquette, Fame, Mallard, Cygnet, Stag	Average 335	30	Guns: 1—12 pdr.; 5—6 pdrs. Torpedo tubes (18 inch): 2.
"C" class.	Falcon, Bittern, Electra, Fairy, Bat, Kestrel, Gipsy, Crane, Recruit, Leven, Star, Vulture, Ostrich, Fawn, Albatross, Osprey, Flying Fish, Velox, Cheerful, Flirt, Leopard, Mermaid, Vigilant, Otter, Racehorse, Thorn, Vixen, Roebuck, Sylvia, Dove, Greyhound, Violet, Bullfinch, Avon, Brazen	Average 335	30	Guns: 1—12 pdr.; 5—6 pdrs. Torpedo tubes (18 inch): 2.

GREAT BRITAIN.

DESTROYERS.

			Displacement (tons)	Speed (knot)	Armament
"B" class.		Albacore Bonetta .	440	30	Guns: 3—12 pdrs. Torpedo tubes (18 inch): 2.
	Kangaroo Peterel Syren Arab Earnest Locust Thrasher Seal Lively Express	Myrmidon Spiteful Success Wolf Griffon Quail Panther Orwell Sprightly	355 to 470	30	Guns: 1—12 pdr.; 5—6 pdrs. Torpedo tubes (18 inch): 2.
"A" class.	Boxer Bruiser Ranger Opossum Sunfish Lightning	Porcupine Surly Conflict Wizard Fervent Zephyr	Average 290	27	Guns: 1—12 pdr.; 5—6 pdrs. Torpedo tubes (18 inch): 2.

BRITISH TORPEDO BOATS.

Summary of 109 Torpedo Boats.

No.	Builders.	Numbers.	Launched.	Displacement.	H.P.	*Max.* speed.	Tubes.	Coal or oil.	Complement.
				tons. *average.*		kts.		tons	
12	*Various*	36—25	1908	260	4000	26	3	24	35
12	*Various*	24—13	1907	260	4000	26	3	24	35
12	*Various*	12—1	1906	225	3750	26	3	20	35
4	White	117—114	1903	205	2900	25	3	23	32
5	Thornycroft	113—109	1902	200	2900	25	3	42	32
4	Thornycroft	99, 98; 107 & 108	1901	178	2850	25	3	20	32
1	Laird	97	1893—94	130	2000	23	3	25	18
3	White	96—94							
3	Thornycroft	93—91							
1	Yarrow	90	1895	100	1430	23	3	18	18
1	Yarrow	88	1894	112	1600	22	3	20	18
6	Yarrow	87—82	1889	85	1100	23	3	20	19
1	White	81	1885	125	...	20	3	35	25
1	Yarrow	80	1887	105	1540	23	5	30	21
1	Yarrow	79	1886	75	1000	22½	3	20	16
16	Yarrow	061—078	1886	75	700	20	5	20	16
19	Thornycroft	060—041	1886	60	700	21	4	16	16
2	Yarrow	039, 040	1885	40	500	17	1	...	15
5	White	038—034	1886	65	950	19	5	...	15

Missing numbers, 56, 62, 75, 87, 89, 100—106, have been lost or struck off list. Most of the old boats have been or are being re-boilered with water tube boilers.

BRITISH SUBMARINES.

75 + *others* BRITISH SUBMARINES.

No.	Type	Date	Displacement	H.P.	*Max.* speed	Fuel	Complement	T. Tubes	*Max.* draug't
			tons		kts.	tons			ft.
8	F 8—1	1913	940/1200	5000/?	20/12	...	...	6	...
1	"S" *Fiat* S.G. type	1913	300/345	...	13/8·5	...	...	2	...
11	E 11—1	1912	725/810	1750/600	16/10	...	...	4	...
6	D 8—D 3	1911	550/600	1750/550	16/10	...	...	3	...
1	D 2	1910							
8	C 38—C 31	1909	280/313	600/200	14/10	15	16	2	12
12	C 30—C 19	1908							
1	D 1	1907	550/600	1200/550	16/9	...	...	3	...
7	C 18—C 12	1907	280/313	600/200	13/8	15	16	2	12
10	C 10—C 1	1906							
10	B 11—B 1	1905	280/313	600/200	13/8	15	16	2	12
8	A 13—A 5	1904	180/207	550/150	11·5/7	7	11	2	11½

E 3, D 5, C 11, B 2, A 7 and *A 3* have been lost.

Submarines under construction included at outbreak of war:

E class up to No. 18.
F class?
Nautilus and *Swordfish*, ocean going, 2,500 h.p.

4 V class (Vickers special) 1,050 tons. 19 kts. surface.
4 W class (Elswick: Laubeuf type).
4 S class (Scott: F.I.A.T. Laurenti type).

Since war began some others may have been commenced, but details are withheld.

GREAT BRITAIN.

MISCELLANEOUS.

			Displacement (tons)	Speed (knots)	Armament
Torpedo Depot Ships.		Woolwich	—	—	Guns: 4—4 inch; 4—12 pdrs.
		St. George	7,700	19.5	—
		Blake Blenheim	9,000	21	Guns: 4—old 6 inch; 16—3 pdrs. Torpedo tubes (14 inch): 2 above water, 2 submerged
		Leander	4,300	—	Guns: 6—3 pdrs.
		Tyne	3,560	11	Guns: 2—24 pdrs.
		Hecla	6.400	12	—
Submarine Depot Ships		**Maidstone**	**3,600**	**14**	—
		Adamant **Alecto**	**935**	**14**	—
		Arrogant	**5,750**	—	—
	(*Astræa class*)	Bonaventure	4,360	19.5	Guns: 2—6 inch; 8—6 pdrs.
		Vulcan	**6,620**	**16**	Guns: 8—4.7 inch; 12—3 pdrs.; 12 Nordenfelts. Torpedo tubes: 2 submerged; 4 above water.
		Thames **Forth**	**4,050**	**17.2**	No guns.

GREAT BRITAIN.

MISCELLANEOUS.

			Displacement (tons)	Speed (knots)	Armament
7 Mine Layers.		Iphigenia . .			
		Thetis . .			
		Apollo . .	3,400		
Apollo class.		Andromache .	to		Guns : 6—6 pdrs.
		Intrepid . .	3,600		
		Latona . .			
		Naiad . .			
3 Repair Ships	(ex *Indiabarrah*)	Cyclops . .	11,000	14	Guns : 6—4 inch.
	(ex *Hampstead*)	Aquarius . .	2,800	10.5	
		Assistance .	9,600	13	Guns : 10—3 pdrs.
1 Hospital Ship.	Maine, *ex* Mediator. Others also chartered temporarily.				

GREAT BRITAIN.
MISCELLANEOUS.

		Displacement (tons)	Speed (knots)	Armament
Surveying Service	Endeavour	1,280	13	Guns: 1—3 pdr.; 2 Maxims.
	Merlin, Fantome	1,070	13.25	
	Mutine	980	13.25	
	Research	520	9.5	Guns: 1 old 7 pdr. M.L.
	Hearty	1,300	—	
	Triton	410	9	Guns: 1 old 7 pdr. M.L.
	Sealark	900	10	
	Esther, Daisy	600		
Miscellaneous Special Service	Fortitude	400		
	Magnet	430	12	Guns: 4—3 pdrs.
	Traveller	700	12	
	Imogene	460	11	
	Sphinx	1,130	12.5	Guns: 1 old 6 inch; 6—old 4 inch.
	Seahorse	670		Guns: 1—12 pdr. howitzer.
	Undine	453	9	
	Vesuvius	245		Torpedo tubes: 2.
Despatch Vessel	Alacrity (ex *Surprise*)	1,650	17.8	Guns: 2—6 pdrs.

GREAT BRITAIN.

MISCELLANEOUS.

		Displacement (tons)	Speed (knots)	Armament
Monitors	Severn Humber Mersey	1,260	11½	Guns: 2—6 inch; 2—4.7 howitzers; 4—3 pdrs. and 6 small.
Royal Yachts	Alexandra Victoria and Albert	2,050 4,700	18.25 20	
Admiralty Yacht	Enchantress	3,470	18	Guns: 4—3 pdrs.

H.M.S. "QUEEN ELIZABETH" (Battleship). *Abrahams*

PLATE II

H.M.S. "AGINCOURT" (Battleship). *Abrahams*

H.M.S. "ERIN" (Battleship). *Cribb*

PLATE IV

H.M.S. "IRON DUKE" (Battleship). *Cribb*

H.M.S. "AUDACIOUS" (Battleship). *Cribb*

PLATE VI

H.M.S. "THUNDERER" (Battleship). *Cribb*

H.M.S. "CONQUEROR" (Battleship). *Cribb*

Plate VIII

H.M.S. "COLOSSUS" (Battleship). *Abrahams*

PLATE IX

H.M.S. "NEPTUNE" (Battleship) *Cribb*

Plate X

H.M.S. "VANGUARD" (Battleship). *Cribb*

H.M.S. "BELLEROPHON" (Battleship). *Cribb*

PLATE XII

H.M.S. "LORD NELSON" (Battleship). *Symonds*

Plate XIII

H.M.S. "DREADNOUGHT" (Battleship). *Cribb*

Plate XIV

H.M.S. "BRITANNIA" (Battleship). *Cribb*

H.M.S. "FORMIDABLE" (Battleship). *Cribb*

PLATE XVI

H.M.S. "QUEEN" (Battleship). *Cribb*

H.M.S. "OCEAN" (Battleship). *Cribb*

PLATE XVIII

H.M.S. "PRINCE GEORGE" (Battleship). *Cribb*

H.M.S. "TIGER" (Battleship).

Plate XX

H.M.S. "QUEEN MARY" (Battle Cruiser). *Cribb*

PLATE XXI

H.M.S. "LION" (Battle Cruiser). *Symonds*

PLATE XXII

H.M.S. "INDEFATIGABLE" (Battle Cruiser). *Cribb*

PLATE XXIII

H.M.S. "AUSTRALIA" (Battle Cruiser). *Cribb*

PLATE XXIV

H.M.S. "INFLEXIBLE" (Battle Cruiser). *Cribb*

H.M.S. "COCHRANE" (Cruiser). *Cribb*

Plate XXVI

H.M.S. "DUKE OF EDINBURGH" (Cruiser) *Cribb*

H.M.S "ANTRIM" (Cruiser). *Cribb*

Plate XXVIII

H.M.S. "KENT" (Cruiser). *Cribb*

H.M.S. "LEVIATHAN" (Cruiser). *Cribb*

PLATE XXX

H.M.S. "EURYALUS" (Cruiser). *Cribb*

PLATE XXXI

H.M.S. "DIADEM" (Cruiser). *Symonds*

PLATE XXXII

H.M.S. "ARETHUSA" (Light Cruiser). *Cribb*

PLATE XXXIII

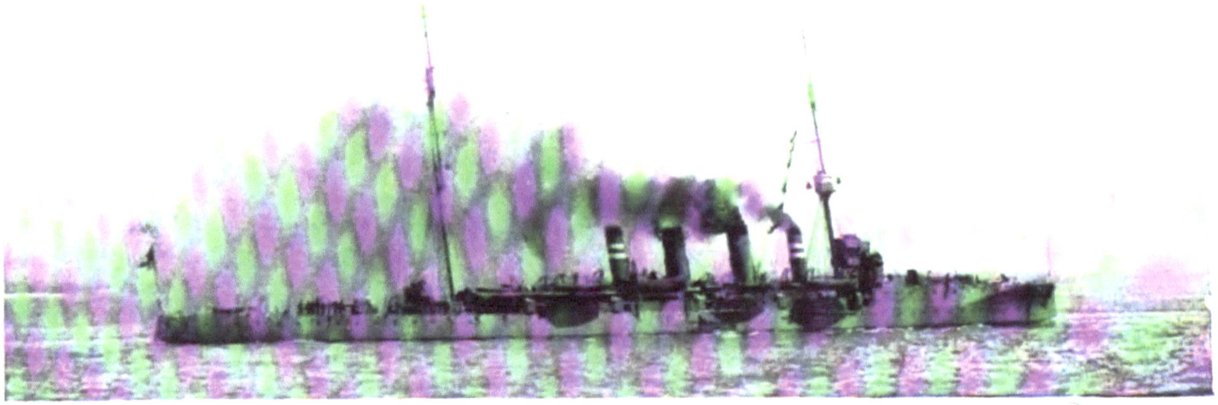

H.M.S. "GLOUCESTER" (Light Cruiser). *Cribb*

Plate XXXIV

H.M.S. "CAROLINE" (Light Cruiser). *Cribb*

PLATE XXXV

H.M.S. "SOUTHAMPTON" (Light Cruiser.) Cribb

Plate XXXVI

H.M.S. "FALMOUTH" (Light Cruiser).

Cribb

Plate XXXVII

H.M.S. "LIVERPOOL" (Light Cruiser). *Cribb*

H.M.S. "FEARLESS" (Light Cruiser). *Cribb*

PLATE XXXIX

H.M.S. "BLONDE" (Light Cruiser). *Cribb*

H.M.S. "BELLONA" (Light Cruiser). *Cribb*

PLATE XLI

H.M.S. "DORIS" (Light Cruiser). *Cribb*

PLATE XLII

H.M.S. "VINDICTIVE" (Light Cruiser). *Cribb*

Plate XLIII

H.M.S. "ASTRÆA" (Light Cruiser). *Cribb*

Plate XLIV

H.M.S. "SKIRMISHER" (Light Cruiser). *Cribb*

PLATE XLV

H.M.S. "AMETHYST" (Light Cruiser). *Cribb*

H.M.S "HUMBER" (Monitor). *Laffayette*

PLATE XLVII

H.M.S. "IPHIGENIA" (Light Cruiser). *Cribb*

PLATE XLVIII

H.M.S. "LEDA" (Gunboat)

Cribb

Plate XLIX

H.M.S. "ANDROMACHE" (Light Cruiser) Apollo Class. *Cribb*

PLATE L.

H.M.S. "WOOLWICH" (Depôt Ship).

Cribb

PLATE LI

H.M.S. "SWIFT" (Flotilla Leader). *Cribb*

H.M.S. "BROKE" (Flotilla Leader).

H.M.S. "LAERTES" (Destroyer).

Cribb

PLATE LV

H.M.S. "HARDY" (Destroyer) Cribb

H.M.S. "PHŒNIX" (Destroyer). Cribb

Plate LVII

H.M.S. "HOPE" (Destroyer). Cribb

H.M.S. "FOXHOUND" (Destroyer). *Cribb*

Plate LIX

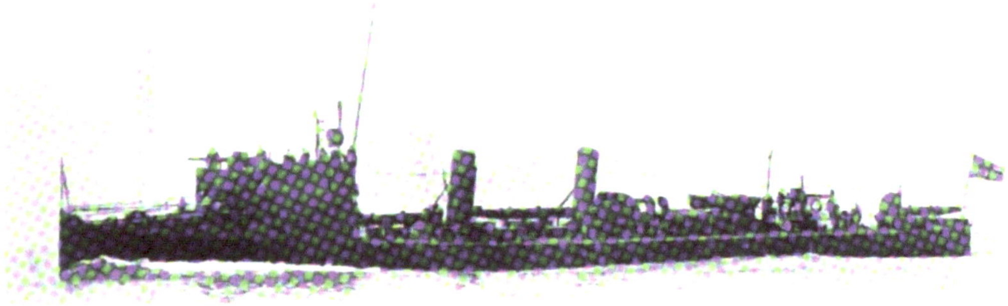

H.M. TORPEDO BOAT 20.

Cribb

Plate LX

H.M.S. "MOHAWK" (Destroyer).

Cribb

H.M.S. "WELLAND" (Destroyer)　　*Cribb*

H.M.S. "BOYNE" (Destroyer). *Cribb*

H.M.S. "ANGLER" (Destroyer) D class.

Cribb

Plate LXIV

H.M.S. "ADAMANT" (Depôt Ship for Submarines). *Symonds*

Plate LXV

H.M.S. "CYCLOPS" (Repair Ship).

H.M.S. "ARK ROYAL" (Special Service Vessel). *Cribb*

H.M. SUBMARINE E 7.

Cribb

PLATE LXVIII

H.M. SUBMARINE D1.

Cribb

Plate LXIX

H.M. SUBMARINE D 7. *Cribb*

Plate LXX

H.M. SUBMARINE C 34. *Cribb*

PLATE LXXI

Cribb

H.M. SUBMARINE C 17.

GREAT BRITAIN.

AUSTRALIAN NAVY.

		Completed	Displacement (tons)	Speed (knots)	Armament	Complement
Dreadnought Cruiser.	Australia . (Sister to the *Indefatigable.*)	1913	19,200	26	Guns: 8—12 inch. 50 cal.; 16—4 inch. Torpedo tubes (21 inch): 3.	800
Protected Cruisers. *Dartmouth class.*	Melbourne . Sydney . Brisbane .	1913 1915	5,600	26	Guns: 8—6 inch; 4—3 pdrs. Torpedo tubes (21 inch): 2.	376
Light Cruiser. *Challenger Class.*	Encounter .	1906	5,900	20	Guns: 11—6 inch (wire); 8—12 pdrs. (12 cwt.); 1—12 pdr. (8 cwt.); 6—3 pdrs.; 2 Maxims. Torpedo tubes (18 inch): 2.	475
	Pioneer					
Torpedo Boat Destroyers.	Parramatta, Yarra, Warrego, Derwent, Swan, Torrens		700	26	Guns: 1—4 inch; 3—12 pdrs. Torpedo tubes (18 inch): 3.	
Miscellaneous	South Australia: Protector		920	14	Guns: 1—old 8 inch; 5—6 inch; 4 Gatlings.	
Miscellaneous	Queensland: Gayundah, Saluma		360	10.5	Guns: 1—old 8 inch; 1—6 inch; 4 Machines.	
One submarine, AE 2 (British E class type). AE 1				16 10		

GREAT BRITAIN

CANADIAN NAVY.

		Completed	Displacement (tons)	Speed (knots)	Armament	Complement
Protected Cruiser. ***Diadem class.***	Niobe . .	1899	11,000	18	Guns: 16—6 inch; 12—12 pdrs. (12 cwt.); 1—12 pdr. (8 cwt.); 12—3 pdrs.; 2 Maxims. Torpedo tubes (18 inch): 2.	677
Light Cruiser ***Apollo class.***	**Rainbow** .		3,600	16	Guns: 2—6 inch (wire); 6—4.7 in.; 8—6 pdrs.; 1—3 pdr. Torpedo tubes (14 inch): 4.	273
Gun Vessels.	**Canada** .		780	17	Guns: 4—3 pdrs.	
	Minto . .		1100	16	Guns: 4—6 pdrs.	
	Curlew . .		400	11	Guns: 3 machine.	
	Constance .		400	11	Guns: 3 machine.	
	Acadia . .		526	10.5	Guns: 6 machine.	

Also 1 Customs Cruiser, *Margaret*; 2 Survey Ships, *Cartier* and *Gauss*; and 2 Vickers-built vessels, *Earl Grey* and *Lady Grey*.

GREAT BRITAIN.

INDIAN MARINE.

		Displacement (tons)	Speed (knots)	Armament
Troop-Ships	Northbrook	5,820	16	Guns: 6—4 inch; 6—3 pdrs.
	Dufferin	7,457	19	Guns: 8—4 inch; 8—3 pdrs.
	Hardinge	6,520	18	Guns: 6—4.7 inch; 6—3 pdrs.; 4 Maxims.
	Minto	930		—
	Dalhousie	1,960	14	Guns: 6—6 pdrs.
	Lawrence	1,154	13	Guns: 4—6 pdrs.; 4 Machine.
	Investigator	1,015	13.5	

GREAT BRITAIN.

LIST OF LINERS, ETC., COMMISSIONED BY H.M. GOVERNMENT

Alsatian
Ambrose
Amsterdam
Anglia
Armadale Castle
Cæsarea
Calgarian
Calyx
Cambria
Campania
Caribbean
Carmania
Caronia
Carron
Cedric
Celtic
Changuinola
City of Belfast
Clacton
Columbella
Digby
Duchess of Devonshire
Duke of Albany
Duke of Cornwall
Edinburgh Castle
Empress
Empress of Asia
Empress of Britain
Empress of Japan
Empress of Russia
Engadine
Eskimo
Fiona
Folkestone
Gazelle
Hazel
Heroic
Hilary
Hildebrand
Himalaya
Hythe
Kinfauns Castle
King Orry
Laconia
Laurentic
Louvain
Macedonia
Mairi
Mantua
Marmora
Motagua
Newmarket
Orama
Oropesa
Orotava
Otranto
Otway
Paris
Partridge
Patia
Patuca
Peel Castle
Ramsey
Reindeer
Riviera
Roedean
Royal Scot
Sarnia
Scotia
Snaefell
Sumatra
Tara
Teutonic
Victorian
Virginian
Woodnut

2. FRANCE.

FRANCE.

DREADNOUGHTS.

		Completed	Displacement (tons)	Speed (knots)	Armament	Complement
New Programme. *Duquesne class.*	Duquesne, Tourville, Lille, Lyon	Building or Projected	29,500	21	Guns : 16—13.4 inch	—
1913 Dreadnoughts. *Normandie class.*	Normandie, Gascogne, Languedoc, Flandres, Bearn	Completing.	25,387	21	Guns : 12—13.4 inch 45 cal. ; 24—5.5 inch, 45 cal. ; 4—3 pdrs. Torpedo tubes (18 inch) : 6.	
1912 Dreadnoughts. *Bretagne class.*	Bretagne, Lorraine, Provence		23,550	20.5	Guns : 10—13.4 in., 45 cal. ; 22—5.5 inch, 45 cal ; 4—3 pdrs. Torpedo tubes (18 inch) : M. '12 : 4.	1109
1910 Dreadnoughts.	France, Courbet	1913	23,467	21	Guns : 12—12 inch, 50 cal. ; 22—5.5 inch, 45 cal. ; 4—3 pdrs. Torpedo tubes (18 inch) : 4.	904
	Jean Bart, Paris	1914				

FRANCE.

BATTLESHIPS (PRE-DREADNOUGHTS).

		Completed	Displacement (tons)	Speed (knots)	Armament	Complement
1907 Battleships. (Semi-Dreadnoughts—*Cuirassés d'Escadre.*) *Danton class.*	Danton . Mirabeau . Diderot . Condorcet . Vergniaud . Voltaire .	1911	18,400	19.4	Guns: 4—12 inch, 50 cal.; 12—9.4 inch, 50 cal; 16—12 pdrs. S.A.; 10—3 pdrs. S.A. Torpedo tubes (M. '04, 18 inch): 2.	681
1903 Battleships. *Vérité class.*	Démocratie . Justice . . Vérité . .	1907 1908	14,900	18	Guns: 4—12 inch, 50 cal.; 10—7.6 inch, 45 cal.; 13—9 pdrs.; 10—3 pdrs. Torpedo tubes (1904 pattern): 4.	793
1901 Battleships. *République class.*	République . Patrie . .	1906	14,865	18	Guns: 4—12 inch, 45 cal.; 18—6.4 inch, 45 cal.; 25—3 pdrs.; 2—1 pdrs. Torpedo tubes (18 inch): 5.	793

FRANCE.

BATTLESHIPS.

		Completed	Displacement (tons)	Speed (knots)	Armament	Complement
1899 Battleship.	**Suffren**	1903	12,750	18	**Guns: 4—12 inch, 45 cal.; 10—6.4 inch, 45 cal.; 8—4 inch; 22—3 pdrs. Torpedo tubes (18 inch): 4.**	730
1897 Battleship.	**Henri IV**	—	8,948	17.5	**Guns: 2—10.8 inch, 45 cal.; 7—5.5 inch, 45 cal.; 12—3 pdrs.; 2—1 pdrs. Torpedo tubes (18 inch): 2.**	**460**
1894 Battleships. *Charlemagne class.*	**Charlemagne** **St. Louis** **Gaulois**	1899 1900 1899	11,260	18	**Guns: 4—12 inch; 10—5.5 inch; 8—4 inch; 20—3 pdrs. Torpedo tubes (18 inch): 4.**	**631**

FRANCE.

OLDER BATTLESHIPS.

	Completed	Displacement (tons)	Speed (knots)	Armament	Complement
Massena .	1898	11,924	18	Guns: 2—12 inch, 40 cal.; 2—10.8 inch, 40 cal.; 8—5.5 inch; 8—4 inch; 12—3 pdrs; 2—9 pdrs. (Boat). Torpedo tubes (18 inch): 4.	617
Carnot . .	1894	12,150	18	Guns: 2—12 inch, 45 cal.; 2—10.8 inch, 45 cal.; 8—5.5 inch; 4—9 pdrs.; 18—3 pdrs. Torpedo tubes (18 inch): 2.	621
Charles Martel	1893	11,882	18	Guns: 2—12 inch; 2—10.8 inch; 8—5.5 inch; 4—9 pdrs.; 20—3 pdrs. Torpedo tubes (18 inch): 2.	
Jauréguiberry	1893	11,900	17.8	Guns: 2—12 inch; 2—10.8 inch; 8—5.5 inch; 16—3 pdrs. Torpedo tubes (18 inch): 2.	607

Note.—One or two old battleships, while for sale, are in temporary use for subsidiary purposes.

FRANCE.

ARMOURED CRUISERS.

		Completed	Displacement (tons)	Speed (knots)	Armament	Complement
1905 Armoured Cruisers.	Edgar Quinet Waldeck Rousseau	1910	14,000	23	Guns (M '02): 14—7.6 inch, 45 cal.; 20—9 pdrs.; 2—1 pdrs. Torpedo tubes (18 inch): 2.	738
1904 Armoured Cruiser.	Jules Michelet	1908	12,000	22	Guns (M. '02): 4—7.6 inch, 45 cal.; 12—6.4 inch, 50 cal.; 24—3 pdrs. Torpedo tubes (18 inch): 4.	724
1903 Armoured Cruiser.	Ernest Renan	1908	13,644	23	Guns: 4—7.6 inch, 45 cal.; 12—6 in., 50 cal.; 16—9 pdrs.; 8—3 pdrs.; 2—1 pdrs. Torpedo tubes (18 inch): 2.	750
1901 Armoured Cruisers. *Gambetta class.*	Jules Ferry . Victor Hugo .	1906	12,416	22	Guns: 4—7.6 inch, 45 cal.; 16—6.4 inch; 45 cal.; 24—3 pdrs. Torpedo tubes (18 inch): 5.	710

FRANCE.

ARMOURED CRUISERS.

		Completed	Displacement (tons)	Speed (knots)	Armament	Complement
1899 Armoured Cruisers. *Aube class.*	Marseillaise . Gloire . . Condé . . Amiral Aube .	1903 1904	10,000	21	Guns: 2—7.6 inch, 40 cal.; 8—6.4 inch, 45 cal.; 6—4 inch; 18—3 pdrs.; 2—9 pdrs. (Boat.) Torpedo tubes (18 inch): 5.	612
1899 Armoured Cruisers. *Dupleix class.*	Dupleix . Kléber . . Desaix . .	1903 1904	7,700	21	Guns: 8—6.4 inch, 45 cal.; 4—4 inch; 10—3 pdrs. Torpedo tubes (18 inch): 2.	520
1897 Armoured Cruisers. *Montcalm class.*	Gueydon . Montcalm . Dupetit Thouars	1902 1903	9,517	21	Guns: 2—7.6 inch; 8—6.4 inch; 4—4 inch; 16—3 pdrs. Torpedo tubes (18 inch): 2.	540 (Flag-ship, 583)
1896 Armoured Cruiser. *Training ship for Midshipmen.*	Jeanne D'Arc	1903	11,270	23	Guns: 2—7.6 inch, 45 cal.; 14—5.5 inch, 45 cal.; 12—3 pdrs.; 8—1 pdrs Torpedo tubes (18 inch): 2.	626
1893 Armoured Cruiser.	Pothuau .	1896	5,360	19	Guns: 2—7.6 inch; 10—5.5 inch; 10—3 pdrs. Torpedo tubes: 4.	463

FRANCE.

PROTECTED CRUISERS.

		Completed	Displacement (tons)	Speed (knots)	Armament	Complement
1897 Protected Cruisers.	Jurien de la Gravière .	1901	5,685	23	Guns: 8—6.4 inch, 45 cal.; 10—3 pdrs. Torpedo tubes: 2.	511
	D'Estrées .	1899	2,460	21	Guns: 2—5.5 inch, 45 cal.; 4—4 in., 45 cal.; 8—3 pdrs. Torpedo tubes: 3.	234
1895 Protected Cruiser.	Guichen .	1902	8,277	23	Guns: 2—6.4 inch, 45 cal; 6—5.5 in., 45 cal.; 10—3 pdrs.; 5—1 pdrs. Torpedo tubes: 2.	625
1894 Protected Cruiser.	D'Entrecasteaux	1898	8,114	19.5	Guns: 2—9.4 inch, 40 cal; 12—5.5 inch, 45 cal.; 12—3 pdrs.; 4 Maxims Torpedo tubes (18 inch): 6.	521
1893 Old Protected Cruisers	Du Chayla .	1898	4,000	19	Guns: 6—6.4 inch, 45 cal.; 4—4 inch; 4—3 pdrs.; 11—1 pdrs. Torpedo tubes (14 inch): 2.	393
	Descartes .	1896	4.000	19.5	Guns: 4—6.4 inch, 45 cal; 10—4 in.; 10—3 pdrs.; 4—1 pdrs. Torpedo tubes (18 inch): 2.	378

FRANCE.

SUBSIDIARY SERVICE CRUISERS.

	Completed	Displacement (tons	Speed (knots)	Armament	Complement
Latouche-Tréville Bruix . . .	1893 1896	4,750	18.5	Guns: 2—7.6 inch; 6—5.5 inch; 6—9 pdrs.; 4—3 pdrs. Torpedo tubes (18 inch): 4.	370
Lavoisier . . .	1898	2,350	20	Guns: 4—5.5 inch, 45 cal.; 2—4 in.; 8—3 pdrs. Torpedo tubes: 2.	248
Friant . .	1895	3,882	19	Guns: 6—6.4 inch, 45 cal.; 4—4 in.; 4—3 pdrs.; 11—1 pdr. Torpedo tubes (14 inch: 2.	393

FRANCE.

SMALL WARSHIPS.

		Completed	Displacement (tons)	Speed (knots)	Armament	Complement
Torpedo Gunboats.	Dunois . . La Hire .	1898 1899	900	23 (designed)	Guns: 6—9 pdrs; 6—3 pdrs. No tubes.	128
	D'Iberville .	1894	925	21.5	Guns: 1—4 inch; 3—9 pdrs.; 6—3 pdrs.	140
Gunboats.	Surprise . Décidée. . Zélée . .		680	13.4	Guns: 2—4 inch; 4—9 pdrs.; 4—1 pdrs.	93
River Gunboats.	Argus . . Vigilante . .		123	13	Guns: 2—3.5 inch; 4—1 pdrs.	
	Doudart de Lagrée Balny . .		213	14.5	Guns: 6—1 pdrs.	
"Avisos."	Kersaint .		1,243	15	Guns: 1—5.5 inch; 5—4 inch; 7 small. No torpedoes.	
	One building .		600		Guns: 1—1 pdr.	45

FRANCE.

DESTROYERS.

		Completed	Displacement (tons)	Speed (knots)	Armament
War boats .	? ?	?	—	—	—
1913 *Bisson class.*	Enseigne Roux . . Mécanicien-principal Lestin Enseigne Gabolde . .	Building	850	30	Guns: 2—4 inch; 4—9 pdrs. Torpedo tubes (18 inch): 4.
	Magon . . . . Mangini . . . Protet . . . . Commandant Lucas .	1915	791		
	Bisson . . . . Renaudin . . .	1913	756		
Henry class.	Enseigne Henry . . Aspirant Herbert . .	1913	450	28.5	Guns: 6—9 pdr. Torpedo tubes (18 inch): 3.

FRANCE.

DESTROYERS.

		Completed	Displacement (tons)	Speed (knots)	Armament
1910 design : *Bouclier class.*	Fourché .	1912	734	31 designed	Guns : 2—4 inch ; 4—9 pdrs. Torpedo tubes (18 inch) : 4.
	Faulx .				
	Cimeterre		720		
	Dague .				
	Dehorter .		756		
	Mehl. .				
	Boutefeu .		703		
	Rivière .		746		
	Bory .				
	Casque .		732	32	
	Bouclier .		692	31	
	Garnier .		732		

FRANCE.

DESTROYERS

		Completed	Displacement (tons)	Speed (knots)	Armament
1906 Design. *Spahi class* . .	Voltigeur . Chasseur . Fantassin . Cavalier .		450	28 designed	Guns : 6—9 pdrs. Torpedo tubes (18 inch) : 3.
	Tirailleur .		420		
	Janissière .		449		
	Hussard . Mameluk .		403		
	Spahi . Carabinier		421		
	Lansquenet		417		
Branlebas class. . .	Branlebas Fanfare .	1908	328	25	Guns : 1—9 pdr. 6—3 pdrs. Torpedo tubes (18 inch) : 2.
Gabion class . . .	Etandard . Fanion . Gabion . Oriflamme Sape . Sabretache	1908	328	28	Guns : 1—9 pdr. 4—6 pdrs. Torpedo tubes (18 (inch : 2.

FRANCE.
DESTROYERS.

Class	Name	Completed	Displacement (tons)	Speed (knots)	Armament
Coutelas class	Cognée	1908	335 normal	28	Guns: 1—9 pdr.; 4—6 pdrs. Torpedo tubes (18 inch): 2.
	Coutelas				
	Carquois				
	Claymore	1906			
	Fleuret	1908			
	Glaive	1910			
	Mortier	1908			
	Obusier				
	Pierrier				
	Poignard	1910			
	Stylet	1906			
	Trombion				
	Trident	1908			
Arc class	Arc	1904	303	28	Guns: 1—9 pdr.; 6—3 pdrs. Torpedo tubes (18 inch): 2.
	Belier				
	Bombarde	1903			
	Catapulte				
	Dard	1904			
	Francisque				
	Fronde	1903			
	Harpon				
	Mousqueton	1904			
	Pistolet	1903			
	Sabre	1904			
	Javeline	1903			
	Mousquet	1903			

FRANCE.

DESTROYERS.

		Completed	Displacement (tons)	Speed (knots)	Armament
Yatagan class.	Epée, Pique, Yatagan	1902	310	26	Guns: 1—9 pdr. 6—3 pdrs. Torpedo tubes (15 inch): 2.
Arbeléte class.	Arbeléte, Arquebuse, Escopette, Pertuisane, Sagaie, Carabine, Epieu, Rapière, Sarabacane	1902	303	28	Guns: 1—9 pdr.; 6—3 pdrs. Torpedo tubes (15 inch): 2.
Durandal class.	Durandal, Fauconneau, Hallebarde	1901	310	28	Guns: 1—9 pdr.; 6—3 pdrs. Torpedo tubes (15 inch): 2.

FRENCH "SEA-GOING TORPEDO BOATS."

17 Torpilleurs de haute mer.

2 screws. 15 inch torpedoes.

All armed with 2—3 pdr. Boilers: Normand, Thornycroft, and Du Temple Guyot.

Totals.	Type.	Date.	Displacement.	I.H.P.	*Max.* Speed.	Coal.	Complement.	Tubes.	*Max.* Draught.
					kts.	tons.			feet.
6	*Audacieux*	1900-1	185	1200	26	25	18	3	9
5	*Borrée*	1900	162	4400	30	18	30	2	8
1	*Forban*	1895	140	3200	31·2	15	28	2	7½
1	*Chevalier*	1893	134	2700	27·2	17	32	2	7
1	*Grenadier*	'92-'93	129	1400	25·21	15½	26	2	8½
2	*Aquilon*	'93-'99	120	2000	26·17	16	34	2	9½
1	*Grondeur*	1890	133	1750	24	20	27	2	5

6 ***Audacieux*** Type: Audacieux, Mistral, Simoun, Sirocco, Trombe, Typhon.

5 ***Borrée*** Type: Tramontane, Borrée, Bourrasque, Cyclone, Rafale.

2 ***Aquilon*** Type: Aquilon, Filibustier.

FRENCH TORPEDO BOATS.

170 Torpilleurs de Defence Mobile (Armed with 2—1 pdrs. and 14 inch torpedoes up to No. 302. Later boats 18 inch torpedoes).

Totals	Numbers	First Begun	Last Completed	Displacement	I.H.P.	*Max.* Speed	Coal	Complement	Tubes	*Max.* Draught
				tons		kts.	tons			
74	369—295	'03	'09	97½	2000	26	...	23	3	10
1	243	'03	'05	90	1500 *t*	25	10	23	3	10
95	292—202	1899	1905	87	1500	25	10	21	2	9
	200—149	1894	1899	80	1100	21	10	16	2	9

Nos. 294, 293 and 243 turbines Of the above boats there have been lost or struck off the list:—Nos. 243, 150, 152, 154, 155—161, 163—169, 172—174, 176, 177—179, 182, 186, 189, 190, 194, 198, 199, 201, 203—205, 207—209, 210, 215, 223, 234, 263, 293, 294; also 166, 180, 152, 271, 290.

FRANCE.

SUBMARINES.

			Completed.	Max. Speed above and below surface. (knots)	Armament.
Improved *Gustave type*		Fulton, Joessel, Laplace, Lagrage, Regnault, Sané, Dupuy de Lôme . .	Building	19—11	Guns : 4—9 pdrs. 8 tubes.
	Diane type.	Diane, Daphne . .		18.5—9.5	Guns : 4—9 pdrs. 10 tubes.
	Bellone type.	Bellone, Hermione, Gorgone		17.5—?	8 tubes (18 inch).
1907-11 French Submarines.	*Clorinde type.*	Amphitrite, Astrée, Arthemis, Aréthuse, Atalante, Amarante Ariane, Andromaque, Clorinde Cornélie	1915	15—9.5	8 tubes (18 inch).
	Gustave Zédé type.	Gustave-Zédé, Néréide .	1915		8 tubes (18 inch).
	Experimental Boats. *Bourdelle type.*	Amiral Bourgois	1912	15—10	7 tubes (18 inch).
	Maurice type .	Charles Brun			6 tubes (18 inch).
	Radiguet type .	Mariotte .	1911		7 tubes (18 inch).
	Huetter type .	Archimède .	1910		

FRANCE.

SUBMARINES.

			Completed	Max. Speed above and below surface (knots).	Armament
1907-11 Submarines.	*Cherbourg Laubœufs.*	Brumaire, Floréal, Frimaire, Fructidor, Germinal, Messidor, Nivôse, Pluviôse, Prairial, Thermidor, Ventose, Euler, Foucault, Franklin . .	1912	12—7.75	7 tubes (18 inch).
	Rochefort Laubœufs	Berthelot, Cugnot, Faraday, Fresnel, Giffard, Montgolfier, Newton, Papin, Volta, Watt			
	Toulon Laubœufs.	Ampère, Arago, Bernouilli, Colomb, Curie, Gay-Lussac, Joule, Leverrier, Monge .		12—7.75	6 tubes (18 inch).
1906-4 Submarines.	*Circe type.*	Circe, Calypso . .	1908	11—8	6 tubes (18 inch).
	Emeraude type.	Eméraude, Opale, Rubis, Saphir, Turquoise, Topaze	1908	12—8	6 tubes (18 inch).
	Aigrette type.	Aigrette, Cigogne . .	1905	10.5—8	4 tubes (18 inch).
	Argonaute type.	Argonaute (ex *Omega*) .	1908	11—8	4 tubes (18 inch).

FRANCE.

SUBMARINES.

			Completed	Max. Speed above and below surface (knots)	Armament
1904-1 Submarines.	*Follet class*	Follet (ex Farfadet)	1902	12—8	4 tubes (18 inch).
	Triton class.	Triton, Silure, Espadon, Sirène	1903	12—8	4 tubes (18 inch).

FRENCH.

MISCELLANEOUS.

			Displacement (tons)	Speed (knots)	Armament
Mine Layers	Special	Pluton Cerbère	566	20	No guns. Carry 60 mines.
	T.B.D's. *Arc class*	Hache Massue Baliste	303	28	Guns: 1—9 pdr.; 6—3 pdrs. Torpedo tubes (18 inch): 2.
	T.B.D. *Arbalète Class*	Flamberge	303	28	Guns: 1—9 pdr.; 6—3 pdrs. Torpedo tubes (15 inch): 2.
	Protected Cruiser	Chateaurenault	8,018	23	Guns: 2—6.4 inch; 6—5.5 inch; 10—3 pdrs.
		Casabianca Cassini	960	21.5	Guns: 1—4 inch; 3—9 pdrs.; 7—3 pdrs.
Submarine Depot Ship		Trehouart	6,535	17	Guns: 2—12 inch, 45 cal. (specially fitted for high angle fire); 8—4 inch; 4—3 pdrs.; 10—1 pdr. Torpedo tubes (18 inch): 2.
Special Aerial Depot Ship		Foudre (Fitted to carry flying machines.)	6,086	19.9	Guns: 8—4 inch; 4—9 pdrs.; 2—3 pdrs.

FRANCE.

MISCELLANEOUS.

		Displacement (tons)	Speed (knots)	Armament
Avisos Transports. (Surveying Vessels.)	Manche Vaucluse	1,600	10–11	Nil.
Transports.	Vinh-Long	5,500	13	Guns: 4—5.5 inch; 2—4 inch; 2—9 pdrs.; 2—3 pdrs.
	Drome	2,200	11	2 small guns.
Special Service. (Pilotage School)	Chamois	425	15	Guns: 4—3 pdrs.
Special Service. (Coastguard Service)	Ibis	254	12	Guns: 2—9 pdrs.

FRENCH "JEAN BART" (Battleship). *Cribb*

PLATE LXXIII

FRENCH "DANTON" (Battleship) *Cribb*

FRENCH "REPUBLIQUE" (Battleship). *Geiser*

Plate LXXV

FRENCH "SUFFREN" (Battleship). *Bar*

FRENCH "CHARLEMAGNE" (Battleship). *Geiser*

PLATE LXXVIII

FRENCH "MARSEILLAISE" (Cruiser). *Cribb*

PLATE LXXVII

FRENCH "CARNOT" (Battleship). *Geiser*

FRENCH "JULES MICHELET" (Cruiser).

Geiser

PLATE LXXX

FRENCH "AMIRAL AUBE" (Cruiser). *Cribb*

FRENCH "POTHUAU" (Cruiser). *Geiser*

PLATE LXXXII

FRENCH "CLAYMORE" (Destroyer). *Cribb*

FRENCH "VENDEMIARE" (Submarine). *Cribb.*

PLATE LXXXIV

FRENCH SUBMARINES AT PORTLAND. *Cribb*

3. RUSSIA.

RUSSIA.

BALTIC FLEET.

		Completed	Displacement (tons)	Speed (knots)	Armament	Complement
1912 Battle- Cruisers. Dreadnoughts).	Borodino . Ismail . . Kinburn . Navarin .	Building.	28,000	—	Guns: 9—14 inch; 20—5.1 inch. Torpedo tubes: —	—
1909 Battle-Cruisers. Dreadnoughts.)	Gangoot . Poltava. . Petropavlovsk Sevastopol .	1914 to 1915	23,300	23	Guns: 12—12 inch, 50 cal.; 16—4.7 inch, 50 cal.; 4—3 pdrs.; 8 mach. Torpedo tubes: 4.	—
1903 Battleships. *Imperator Pavel class.*	Imperator Pavel Andrei Pervoswanni	1911	17,400	18	Guns: 4—12 inch, 40 cal.; 14—8 inch, 50 cal.; 12—4.7 inch; 4—3 pdrs.; 8 machine. Torpedo tubes (18 inch): 3.	800
1902 Battleship.	Slava . .	1905	13,500	18	Guns: 4—12 inch, 40 cal.; 12—6 in., 45 cal.; 20—12 pdrs.; 4—3 pdrs.; 8 Machine. Torpedo tubes (18 inch): 4.	750

RUSSIA.

BALTIC FLEET.

		Completed	Displacement (tons)	Speed (knots)	Armament	Complement
1899 Battleship	Tsessarevitch	1903	13,380	18	Guns: 4—12 inch, 40 cal.; 12—6 in., 45 cal.; 20—12 pdrs.; 20—3 pdrs. 6 Machine. Torpedo tubes (18 inch): 4.	750
1905 Armoured Cruiser.	Rurik . .	1908	15,000	21	Guns: 4—10 inch, 50 cal.; 8—8 in., 50 cal.; 20—4.7 inch; 14 smaller. Torpedo tubes (18 inch): 2.	899
1905 Armoured Cruisers. *Bayan class.*	Admiral Makaroff Bayan . . [illegible]	1907 1910	7,775	22	Guns: 2—8 inch, 45 cal.; 8—6 inch, 45 cal.; 22—12 pdrs. 4—3 pdrs. Torpedo tubes: 2.	568
1898 Armoured Cruiser.	Gromoboi .	1901	13,200	20	Guns: 4—8 inch, 45 cal.; 22—6 inch, 45 cal.; 4—3 pdrs. Torpedo tubes: 4.	800
1893 Armoured Cruiser.	Rossiya. .	1898	12,500	20	Guns: 4—8 inch, 45 cal.; 22—6 inch, 45 cal.; 15-12 pdrs. 36 smaller Q.F. Torpedo tubes: 6.	735

RUSSIA.

BALTIC FLEET.

		Completed	Displacement (tons)	Speed (knots)	Armament	Complement
1913 Cruisers.	Swietlana . . Greig . . . Bootakof . . Spiridof . .	Building.	6,750		Guns : 12—6 inch. Torpedo tubes (18 inch) : —	—
	*Mooraviev Amursky *Nevelskoy . .	1914-15	4,500	27.5	Guns : 8—5 inch ; 4—2½ inch. Torpedo tubes (18 inch) : 2.	—
1901 Cruiser.	Askold . .	1903	6,500	23	Guns : 12—6 inch, 45 cal. ; 12—12 pdrs. ; 8—3 pdrs. ; 2—1 pdr. Torpedo tubes (18 inch) : 6.	580

*Completing in Germany about the outbreak of war, and reported seized for the German Navy.

RUSSIA.

BALTIC FLEET.

		Completed	Displacement (tons)	Speed (knots)	Armament	Complement
1899 Cruisers.	Bogatyr . Oleg . .		6,650	23	Guns: 12—6 inch. 45 cal.; 12—12 pdrs.; 8—3 pdrs.; 2—1 pdr. Torpedo tubes (18 inch): 4.	573
	Diana . Aurora .		7,600	20	Guns: 10—6 inch. 45 cal.; 22—12 pdrs.; 8 small. Torpedo tubes (18 inch): 4.	570

RUSSIA.

BALTIC FLEET TORPEDO BOAT DESTROYERS.

36 new boats of *Novik* type (building, completing or projected): **Axard, Avtroil, Briatschislaw, Cap. 2r. Kern, Cap. 2r. Kinsbergen, Cap. 2r. Konon-Zotof, Cap. 2r. Krown, Cap. 2r. Tsylmettiev, Chios, Desna, Gavriil, Grom, Grengamn, Gogland, Koolm, Konstantin, Lieut. Belle, Lieut. Doobassov, Lieut. Ilin, Lieut. Lombardo, Lietoon, Michail, Orphey, Patrass, Pobieditel, Priamishaw, Rymnick, Ssamson, Ssmolensk, Ssokol, Stirssooden, Tenedoss, Theodor Sstratilat, Wladimir, Zabiaka.**

1 **Novik.**

4 "*Kondratenko*" *type* named **Kondratenko, Sibriski-Strelok, Oobotnik, Pogranitschiek.**

4 "*Emir Bucharski*" *type* named **Emir Bucharski, Dobrovoletz, Finn, Moskwityanin.**

4 "*Amuretz*" *type* named **Amuretz, Gaidamak, Vsadnik, Ussurietz.**

8 "*Strashni*" *type* named **Donskoï-Kasack, Kasanets, Storogouchi, Strashni, Turkmenets-Stavropolsky, Ukraïna, Voiskovoi, Zabaikaletz.** (3 funnels).

10 "*Boyevoi*" *type* named **Boyevoi, Bditelni, Burni, Leit. Sergieff, Kapitan-Jurassowski, Insch. Mech. Swereff, Insch. Mech. Dmitrieff, Vnimatelni, Vnuschitelni, Vinoslivi.**

"*Dyelni*" *type* named **Dyelni, Dostoini, Dyejatelni, Rastoropni, Rasyaschtschi, Silni, Stroiny, Storoschevoi.**

6 "*Bravy*" *type* named **Bravy, Boiky, Bodry, Groeny, Gromyastchy, Vidny** (with bow tube).

3 "*Bestraschni*" *type* named **Bestraschni, Besposchtschadni, Beschumni.**

"*Lovcki*" *type* named **Legki, Letutschi, Lowki, Leit. Bourakoff, Likoi, Iskussni, Ispolnetelni, Krepki, Metki, Molodetski, Moschtschny.**

"*Twerdi*" *type* named **Leit. Malejeff, Insch. Mech. Anastassoff, Twerdi, Totschni, Trevoschni.**

"*Smyeli*" *class* named **Smyely, Serdity, Statny, Skory, Retivy, Ryaevy, Ryany, Prosorlivy.**

2 "*Grosovoi*" *type* named **Grosovoi, Vlastni.**

6 "*Pritki*" *type* named **Pritky** (ex *Sokol*), **Protschny, Podvischny, Pronsitelny, Poraschamechtshy, Posluschny.**

RUSSIA.

BALTIC FLEET TORPEDO BOATS.

16 Russian (Baltic) + 2 (Siberian) old *Schichau* Torpedo Boats.

No.	Type	Date	Displacement	H.P.	Max. speed	T Tubes
			tons		kts.	
1	*Lastochka* (Y)	1904	200	(*t*)	30	3
2	213, 212 (Y)	1902	186	2500	24	3
8	214 - 220, 222 (N)	1902-3	150	1300	30	
5	142, 140, 134, 129, 128 (N)	1894-8	120	2000	21	3

(Y) = Yarrow. (N) = Normand. (*t*) = turbine.

Siberian Flotilla.

old *Schichau* (206 and 205), dating from 1889. 85 ton type.

RUSSIA.

BALTIC FLEET SUBMARINES

13 + *13* Baltic Submarines.

No.	Type	Date	Displacement	H.P.	Max. speed	T. tubes
					kts.	
12	*New (A)*	*bldg.*	...	...	...	...
1	*New boat*	*bldg.*	500	...	16/9	2
1	*Akula (B)*	1909	370	...	13	2
1	*Poschtovi*	1908	134	...	[illegible]/6	4
1	*Minoga (B)*	1908	120	...	...	2
4	*Alligator (L)*	1908	450/500	...	15/12	4
2	*Makrel (B)*	1907	150/200	...	8/6	4
1	*Sig (L)*	1905	135/175	...	10/7	3
3	*Sterliad (H)*	1904	115/150	160	9½/7	2

(B) = Bubnoff. (L) = Lake. (H) = Holland.

The 12 new boats: **Jaguar, Kaguar, Leopard, Lwiza, Pantera, Ryss, Tigr, Tur, Baro, Gepard, Vyopr, Wolk.**

Makrel class, named **Makrel, Okun.**

Alligator class, named **Alligator, Drakon, Kaiman, Krokodil.**

Sterliad class, named **Sterliad, Byaluga, Peskar.**

12 + *6* Siberian Flotilla.

No.	Type	Date	Displacement	H.P.	Max. speed	T. tubes
6	*New*	*bldg.*	...	...	...	...
5	*Ossetyr (L)*	1904-5	135/175	...	10/7	3
2	*Som (H)*	1904	115/150	160	9½/7	2
4	*Makrel (B)*	1904	150/200	...	8/6	4
1	*Delfin*	1903	115/150	120	8½/6	2

(B) = Bubnoff. (L) = Lake. (H) = Holland.

Ossetyr (ex Protector) type, named **Ossetyr, Bytschok, Kefal, Paltus, Plotwa.**

Som type, named **Som, Schtschuka.**

Makrel type, named **Graf Sheremetyeff, Nalim, Kassatka, Skat.**

RUSSIA.

BALTIC FLEET (MISCELLANEOUS).

		Displacement (tons)	Speed (knots)	Armament
Gunnery School Ships.	Imperator Aleksandr II.	9,900	15.9	Guns: 2—12 inch, 30 cal.; 1—8 inch; 10—6 inch; 4—4.7 inch; 4—3 pdrs.
	Petr Veliky . . .	9,665	—	Guns: 4—8 inch; 10—6 inch; 12—12 pdrs.; 4—6 pdrs.; 4—3 pdrs.; 2 Machine.
River Gunboats.	Grosa, Schkwal, Schtorm, Smertch, Tifun, Uragan, Wiehr, Wyuga	950	11	Guns: 2—6 inch; 4—12 pdrs.
	Ardagan, Kars . .	630	14	Guns: 2—4.7 inch; 4—12 pdrs.
	Giliak, Korietz, Sivoutch, Bohr . . .	960	12	Guns: 2—4.7 inch; 4—12 pdrs.; 3 Machine.
	Khivinetz . . .	1,316	13	Guns: 2—4.7 inch; 8—12 pdrs.; 2 Machine.
	Khrabry . . .	1,700	14.5	Guns: 2—8 inch, 45 cal.; 1—6 inch, 45 cal.; 5 small. Torpedo tubes (14 inch): 1.
	Mandjur . . .	1,213	13	Guns: 2—8 inch; 1—6 inch; 4—4 inch.
Despatch Vessels. . . .	Posadnik . .	411	21	Guns: 3—3 pdrs.; 3—1 pdrs.
	Voevoda . .	411		

RUSSIA.

BALTIC FLEET (MISCELLANEOUS)

		Displacement (tons)	Speed (knots)	Armament
Mine Layers and Special Service Ships	Yenesei, Amur	3,600	17	Guns: 5—4.7 inch; 2—12 pdrs.
	Volga	1,452	—	
	Bakane, Mogutchi	750	—	Guns: 4—1 pdrs.
	Onega, Narova	4,600	—	—
	Ladogo	6,200	10	—
Mine School Ships	Nikolaieff	14,000	13	
	Dwina	6,700	—	
Training Ships	Okean	12,000	18	
	Wierny, Voin	1,280	9	Guns: 4—12 pdrs.
Special Colliers	Suchon, Petschora, Mesen	2,000		
Imperial Yachts	Alexandria	500	14.5	
	Standart	5,557	21	Guns: 8—3 pdrs.
	Polarnaia-Sviezda	3,640	18.8	Guns: 6—12 pdrs.

RUSSIA.

BALTIC FLEET (MISCELLANEOUS).

		Displacement (tons)	Speed (knots)
Ice-Breakers	Ermack	8,000	—
	Nadiejny	1,500	14.75
Special Service Vessels	Oka, Kama	—	—
	Riga	14,500	—
	Angara	11,700	20
	Xenia	6,000	—
	Borgo, Lachta	5,000	—
	Anadyr	10,000	—
	Taimyr, Waigatsch	1,300	—
Submarine Tenders	Chabarovsk	2,800	12.5
	Europa	3,200	13
Mine Sweeper	Minrep	—	—

RUSSIA.

BLACK SEA FLEET.

		Completed	Displacement (tons)	Speed (knots)	Armament	Complement
1911 Dreadnoughts.	Ekaterina II . Imperatriza M. Alexander III	1914	22,500	21	Guns : 10—12 inch ; 20—5 inch ; 4—2½ inch ; 4—3 pdrs. Torpedo tubes : 4.	—
1903 Battleships. *Ievstafi class.*	Ievstafi Ioann Zlatoust	1910	12,800	16	Guns : 4—12 inch, 40 cal. ; 4—8 in., 50 cal. ; 12—6 in., 45 cal. ; 14—12 pdrs. ; 2 Machine. Torpedo tubes : 4.	731
1898 Battleship. *Ievstafi class.*	Pantelimon .	1903	12,800	16	Guns : 4—12 inch, 40 cal. ; 16—6 in., 45 cal. ; 14—12 pdrs. 2 Machine. Torpedo tubes : 4.	731
1896 Battleship.	Rostislav .	1898	9,000	16	Guns : 4—10 inch, 45 cal. ; 8—6 inch, 45 cal. ; 12—3 pdrs. ; 4—1 pdrs. Torpedo tubes (18 inch) : 6.	631
1893 Battleship.	Tri Sviatitelia	1897-8	12,540	18	Guns : 4—12 inch, 35 cal. ; 10—6 in., 45 cal. ; 4—4.7 in. Torpedo tubes (18 inch) : 2.	731
Old Battleship.	Georgi Pobiedonosets	1894	10,250	16.5	Guns : 6—old 12 in.; 7—6 inch ; 8—12 pdrs. Torpedo tubes : 7.	560

RUSSIA.

BLACK SEA FLEET.

		Completed	Displacement (tons)	Speed (knots)	Armament	Complement
Protected Cruisers.	Kagoul (ex Otchakov) Pamiat Merkooria (ex Kagoul)		6,750	23	Guns: 12—6 inch; 12—12 pdrs.; 7 small. Torpedo tubes (18 inch): 2.	573
Armed Yacht.	Almaz . .	1904	3,300	19	Guns: 4—12 pdrs.; 4—3 pdrs.	275
Mine Layers.	Proot (ex Moscwa Dunai . .	— —	5,500 1,360	12 14	Guns: 8—3 pdrs.; 2—1 pdrs.	
Torpedo School Training Ship.	Beresan .	—	3,050	—		
Gunboats.	Donets . . Uralets . Terets . . Koubanets .	1887 refitted 1904-6	1,200	—	Guns: 2—6 inch, 45 cal.; 1—4.7 inch, 45 cal.; 2—12 pdrs.; 4—3 pdrs.; 2—1 pdr.	
Repair and Depot Ship.	Kronstadt .	—	16,000	13		

RUSSIA.

BLACK SEA FLEET.

26 Torpedo Boat Destroyers.

No.	Type	Date	Displacement	H.P.	Speed	Complement	T. tubes
			tons		kts.		
9	*Bistry*	'11–'14	1050	22,500	33 (t)	93	3
4	*Baranoff*	'05–'08	600	6500	26	85	3
9	*Zavidny*	'03–'06	350	5500	27	64	2
4	*Strogy*	'00–'02	240	3800	26	56	2

9 boats (1912-13), named:—***Bistry, Bespokoiny, Dersky, Gnievny, Gromky, Pospyechny, Pronsitelny, Pulky, Stschastliry.***

4 boats (1907). ***Kap. Leit. Baranoff, Leit. Shestakof, Kapitan Saken*** (*ex Pushkin*), ***Leit. Zatzarennyi.***

9 boats (Laird type). ***Zavidny, Zavetny, Jeevoi, Jeevootchy, Jootky, Jarky, Zvonky, Leit. Pushtscheen*** (*ex Sadorny*), ***Zorky*** (1905).

4 boats (early Yarrow type). ***Strogy, Smetlivy, Stremitelny, Svirepy*** (1901).

RUSSIA.

BLACK SEA FLEET.

10 Torpedo Boats.

No.	Type	Date	Displacement	H.P.	*Max.* speed	T. tubes
			tons		kts.	
4	273–270 (N)	'95–'96	120	2000	21	3
6	256, 253 & 252, 263, 260, 259 (S)	'95–'86	104	1000	17	2

(N) = Normand. (S) = Schichau.

There are also a certain number of very old second class boats.

11 Submarines

No.	Type	Date	Displacement	H.P.	*Max.* speed	T. tubes
			tons		kts.	
6	*Kaschalot* (B)	'09–'12	450	...		...
1	*Krab*	1910	400/500	...	...	2
2	*Karp* (G)	1907	$\frac{200}{240}$	...	$\frac{10}{8}$	1
1	*Sudak* (H)	1907	$\frac{115}{150}$	...	$\frac{9}{6}$	2
1	*Losos* (H)	1907	$\frac{120}{150}$	...	$\frac{9}{7}$	2

(B) = Bubnolf. (G) = Germania (Krupp). (H) = Holland

6 *Kaschalot* class, named:—***Kaschalot, Kit, Morsch, Narval, Nerpa, Tyulen***— ***Krab*** (mine)

Karp class, named ***Karp, Karas.***

RUSSIAN "IMPERATOR ALEKSANDR III" (Battleship).

PLATE LXXXVI

RUSSIAN "OLEG" (Cruiser). *Cribb*

RUSSIAN "ROSSIYA" (Cruiser). *Cribb*

Plate LXXXVIII

RUSSIAN "RURIK" (Cruiser).

Cribb

4. JAPAN.

JAPAN.

BATTLESHIPS.

		Completed	Displacement (tons)	Speed (knots)	Armament	Complement
1912	Fu-so . . " B " . . " C " . . " D " . .	1915 1916 1917	30,000	40,000 H.P. kts.	Guns: 12—14 inch, 45 cal.; 16—6 in., 50 cal. Torpedo tubes: —	—
1909	Kawachi . Settsu . .	1912 1913	21,420	20	Guns: 12—12 inch, 45 cal.; 10—6 in., 50 cal.; 8—4 7 inch, 50 cal.; 12—14 pdrs.; 4 Mach. Torpedo tubes (18 inch): 5.	960
1905 Semi-Dreadnought. *Satsuma type.*	Aki . .	1911	19,800	20	Guns: 4—12 inch, 45 cal.; 12—10 inch, 45 cal.; 8—6 inch, 45 cal.; 16—12 pdrs.; 4 Mach. Torpedo tubes (18 inch): 5.	940
	Satsuma .	1910	19,370	18.25	Guns: 4—12 inch, 45 cal.; 12—10 inch, 45 cal.; 12—4.7 inch, 45 cal.; 4—12 pdrs.; 4 Machine. Torpedo tubes (18 inch): 5.	

JAPAN.

BATTLESHIPS.

		Completed	Displacement (tons)	Speed (knots)	Armament	Complement
1904. *Kashima class*	Kashima . . Katori . .	1906	16,400 15,950	18.5	Guns: 4—12 inch, 45 cal.; 4—10 in., 45 cal.; 12—6 in., 45 cal.; 12—14 pdrs.; 6—3 pdrs.; 6 Pompoms. Torpedo tubes (18 inch): 5.	864
1899.	Mikasa . .	1902	15,200	18	Guns: 4—12 inch, 45 cal.; 14—6 in., 45 cal.; 20—12 pdrs.; 8—3 pdrs.; 4—2½ pdrs.; 8 Maxims. Torpedo tubes (18 inch): 4.	756
1898.	Hizen . .	1902 (Reconstructed, 1907)	12,700	18	Guns: 4—12 inch, 40 cal.; 12—6 in., 45 cal.; 20—12 pdrs.; 20—3 pdrs. 6—1 pdr. Torpedo tubes (18 inch): 4.	750
1897. *Shikishima class.*	Asahi . . Shikishima .	1901	15,200	18	Guns: 4—12 inch, 40 cal.; 14—6 in., 40 cal.; 20—12 pdrs.; 8—3 pdrs. 6—2½ pdrs. Torpedo tubes (18 inch): 4 in Asahi; 5 in Shikishima.	741

JAPAN.

BATTLE-CRUISERS.

		Completed	Displacement (tons)	Speed (knots)	Armament	Complement
1911. . .	Kongo . . Hi-yei . . Haruna . Kirishima .	1913 1914 1915	27,500	27	Guns: 8—14 inch, 45 cal.; 16—6 in., 50 cal.; 16—14 pdrs. Torpedo tubes (21 inch): 8.	1,100
1905. . .	Kurama . Ibuki . .	1910 1909	14,600	21.25	Guns: 4—12 inch, 45 cal.; 8—8 in., 45 cal.; 14—4.7 inch; 4 Machine. Torpedo tubes (18 inch): 2.	817
1905. . .	Tsukuba . Ikoma . .	1907	13,750	20.5	Guns: 4—12 inch, 45 cal.; 12—6 in., 45 cal.; 12—4.7 inch; 2—14 pdrs.; 3—3 pdrs.; 6 Machine. Torpedo tubes (18 inch): 3.	817

JAPAN.

FIRST CLASS COAST DEFENCE SHIPS.

		Completed	Displacement (tons)	Speed (knots)	Armament	Complement
1900.	Iwami	1904 Reconstructed, 1906–7	13,566	18	Guns: 4—12 inch, 45 cal.; 6—8 in., 45 cal.; 20—12 pdrs.; 20—3 pdrs. 8—1 pdr. Torpedo tubes (18 inch): 2.	750
1895.	Sagami Suwo	1901 Reconstructed, 1908–9	13,500	19	Guns: 4—10 inch, 45 cal.; 10—6 in., 45 cal.; 16—12 pdrs.; 26 smaller, Q.F. and Machines Torpedo tubes (18 inch): 2.	732
Coast Defence Battleship.	Fuji	1897	12,300	15 (18.5 nominal)	Guns: 4—12 inch, 40 cal.; 10—6 in., 40 cal.; 16—12 pdrs.; 4—2½ pdrs. Torpedo tubes (18 inch): 5.	600
1893-1887 Old Coast Defence Battleships.	Tango	1900	11,000	17	Guns: 4—12 inch, 35 cal.; 12—6 in., 45 cal.; 14 smaller	750
	Iki		9,670	15.9	Guns: 2—old 12 in. 4—old 9 in.; 7—6 in.; 20 light guns. Torpedo tubes (14 inch): 6.	

JAPAN.

FIRST CLASS CRUISERS.

		Completed	Displacement (tons)	Speed (knots)	Armament	Complement
	Nisshin Kasuga	1904	7,750	20	Guns: 4—8 inch, 45 cal.; 14—6 inch, 45 cal.; 10—12 pdrs.; 6—3 pdrs; 2 Maxims; 2 Field guns. Torpedo tubes: 4.	525
	Aso	1903	7,800	21	Guns: 2—8 inch, 45 cal; 8—6 inch, 45 cal.; 20—12 pdrs.; 7—3 pdrs. Torpedo tubes: 2.	500
1898. *Iwate class.*	Idzumo Iwate	1901	9,800	20.75	Guns: 4—8 inch, 40 cal.; 14—6 inch, 40 cal.; 12—12 pdrs.; 8—2½ pdrs.; 4 small. Torpedo tubes: 4	483
1897. *Asama type.*	Yakumo	1901	9,850	20	Guns; 4—8 inch, 40 cal.; 12—6 inch, 40 cal.; 12—12 pdrs.; 7—2½ pdrs. Torpedo tubes (18 inch): 5.	500
	Azuma	1901	9,456	20	Guns: 4—8 inch, 40 cal.; 12—6 inch, 40 cal.; 12—12 pdrs.; 12—3 pdrs. Torpedo tubes: 5.	482
1897. *Asama class.*	Asama Tokiwa		9,750	21.5	Guns: 4—8 inch, 40 cal.; 14—6 inch, 40 cal.; 12—12 pdrs.; 7—2½ pdrs. Torpedo tubes (18 inch): 5.	500

JAPAN.

SECOND CLASS CRUISERS.

	Completed	Displacement (tons)	Speed (knots)	Armament	Complement
Hirato . Yahagi Chikuma	1912	5,040	23 (Designed, =26)	Guns: 6—6 inch, 50 cal.; 8—12 pdrs.; 2 Field guns. Torpedo tubes (18 inch): 2.	392
Tone .	1909	4,105	23	Guns: 2—6 inch, 50 cal.; 12—4.7 inch; 2 Field guns. Torpedo tubes (18 inch): 2.	392
Otowa .	1904	3,000	21	Guns: 2—6 inch, 50 cal.; 6—4.7 inch; 4—12 pdrs.; 2 Machine. No torpedo tubes.	312
Niitaka Tsushima	1904	3,420	20	Guns: 6—6 inch, 40 cal.; 10—12 pdrs.; 4—2½ pdrs. No torpedo tubes.	320
Kasagi Chitose .	1899	4,760	22.5	Guns: 2—8 inch, 40 cal.; 10—4.7 inch; 12—12 pdrs.; 6—2½ pdrs. Torpedo tubes: 4.	405
Akashi .	1898	2,700	20	Guns: 2—6 inch, 40 cal.; 6—4.7 inch, 40 cal.; 12—3 pdrs.; 4 Nordenfelts. Torpedo tubes: 2.	275
Soya .	1901	6,500	23	Guns: 12—6 inch, 45 cal.; 12—12 pdrs.; 6—3 pdrs. Torpedo tubes (18 inch): 4.	571
Tsugaru	1902 (refit 1907)	6,630	20	Guns: 8—6 inch, 45 cal.; 22—12 pdrs.; 8 small Q.F. Torpedo tubes: 4.	570
Suma .	1898	2,657	20	Guns: 2—6 inch, 40 cal; 6—4.7 in., 40 cal.; 12—3 pdrs. Torpedo tubes: 2.	275

JAPAN.

SECOND CLASS COAST DEFENCE SHIPS.

		Completed	Displacement (tons)	Speed (knots)	Armament	Complement
1893-1887 Old Coast Defence Battleships.	Okinoshima Mishima .	1898 1896	4,200	16	Guns: 3—10 inch, 45 cal.; 4—4.7 inch; 10—3 pdrs.; 12—1 pdrs. Torpedo tubes (18 inch): 4.	318
1889 Coast Defence Ship.	Chiyoda .		2,450	19	Guns: 10—4.7 inch, 40 cal.; 15—3 pdrs.; 3 Gatlings. Torpedo tubes: 3.	350
1892 Protected Cruiser.	Akitsushima	1894	315	16	Guns: 4—6 inch; 6—4.7 inch; 10—3 pdrs. Torpedo tubes (14 in.): 4.	330
Old Light Cruisers.	Itsukushima Hashidate .	1891 1894	4,277	16.7	Guns: 1—12.6 inch; 11—4.7 inch; 5—12 pdrs.; 11—3 pdrs.; 2 Max. Torpedo tubes (14 inch): 4.	360
Gunboats. .	Mogami	1909	1,329	23	Guns: 2—4.7 inch; 4—12 pdrs.	177
	Yodo .		1,230	22	Guns: 2—4.7 inch; 4—12 pdrs. Torpedo tubes (18 inch): 3.	125
	Chihaya .	1903	1,250	21		
	Tatsuta .	1896	875	21	Guns: 2—4.7 inch; 4—3 pdrs. Torpedo tubes (14 inch): 2.	100

JAPAN.

54 + 2 TORPEDO BOAT DESTROYERS.

Totals.	Class.	Date.	Displacement.	H.P.	Max. Speed.	Complement.	Tubes. 18 in.
2	*New* (Y)	'12–'14	838	22,000	... *t*	...	...
2	*Sakura*	'11–'13	600	18,000	33*t*	...	2
1	*Umikaje*	'09–'11	1100	21,000	35*t*	123	3
1		'08–'10					
37	*Arare*	'05–'08	380	6000	29	70	2
4	*Asashiho* (T)	'01–'04	365	6000	31	60	2
5	*Akebono* (Y)	'97–'00	306	6000	31	55	2
6	*Kagero* (T)	'97–'00	275	5400	30	54	2
3	*Yamahiko* (Y)	'00–'02	303	5000	27	...	2

(T) = Thornycroft. (Y) = Yarrow type. *t* = Turbine.

Boats over 1000 tons are officially rated first class; 1000—600, second class; and under 600 tons, third class.

2 *Sakura* class:—**Sakura, Tachibana.**
Umikaje and **Yamikaje.**
The 37 *Arare* type are named:—**Arare, Ariake, Asakase, Asatsuyu, Fubuki, Fumitzuki, Hayakase, Hatsuharu, Hatsushimo, Hibiki, Hatsuyuki, Harukaze, Kisaragi, Kamikaze, Kikutsuki, Minatsuki, Mikazuki, Matsukase, Nagatsuki, Nowake, Nenohi, Oite, Shiratsuyu, Shirayuki, Shirotaye, Satsuki, Shigure, Utsuki, Ushio, Wakaba, Yugure, Yayoi, Yudachi, Yunagi.** Also **Ajanami, Iranami, Uranami**
The 4 *Asashiho* type:—**Asashiho, Asagiri, Murasame, Shirakumo.**
The 5 *Akebono* type:—**Akebono, Oboro.**
The 6 *Kagero* type:—**Kagero, Murakumo, Shinonome, Shiranui, Usugumo, Yugiri.**

JAPAN.

TORPEDO FLOTILLA AND SUBMARINES.

55 Torpedo Boats.

No.	Type.	Date.	Displacement. *about*	*Maximum* Speed.	Complement.	Tubes.
			Tons.	kts.		
	First class					
1[illegible]	Aotaka	'03-'05	120	27		3
5	Hayabusa	'98-'01	150	29		3
	Second class					
14	75—62 (Y)	'02-'01	82	24	20	3
11	49—31 (S)	'00	82	24	...	3
4	43—39 (Y)	'00	110	24	...	3
2	30—29 (N)	'98	87	26	...	3
8	*various*	'98	55	21	...	2

Many of the boats were built in Japan, following with slight variations, Yarrow (Y), Schichau (S), or (N) Normand types.

The *Aotaka* class are named ***Aotaka, Hato, Hibari, Kari, Kiji, Tsubame, Sagi, Uzura, Kumome, Hashitaka,*** and ***Otori.***

Hayabusa are the ***Hayabusa, Kasasagi, Shirataka, Manazuru, Chidori.***

15 + *2* Submarines.

No.	Type	Date	Displacement	H.P.	*Max.* speed	T. Tubes
			tons		kts.	
2	*New* Schneider (16, 17)	'12—'14	520 / ·	2100 / ·	17.5 / 9	6
3	13, 14, 15	'11—'12	350	...	...	...
3	10, 11, 12 }	'09—'11	...	...	...	...
2	Vickers C type (8, 9) }	'07—'08	314	600 / 180	13 / 8	2
2	Japanese (6—7)	'06	85			1
5	Elementary Holland (1—5)	'04	125	160 / 70	9 / 7	1

The 5 Hollands were imported from the U.S.A. about the time of the war with Russia.

JAPAN.

MISCELLANEOUS.

		Displacement (tons)	Speed (knots)	Armament
Torpedo Depot Ships. .	Karasaki (ex *Ekaterinoslav*	10,500	13	—
	Toyohashi	4,200	—	—
Despatch Vessels. . .	Manshu (ex *Manchuria*) .		14	—
	Matsue			
Hospital Ships. . .	Hakuai	2,400	—	
	Kosai			
Engineers' Training Ship.	Yuragawa Maru . .			
River Gunboats. . .	Uji	620	13	Guns : 4—12 pdrs.
	Fushimi	180	14	Guns : 2—6 pdrs. ; 4 Maxims.
	Toba	250		
	Sumida	132	15	

JAPANESE "KATORI" (Battleship). *Symonds*

PLATE XC

JAPANESE "KONGO" (Battle Cruiser). *Cribb*

JAPANESE "IKOMA" (Cruiser). *Cribb*

PLATE XCII

JAPANESE "KURAMA" (Cruiser).

Silk

II. THE ENEMY ALLIED POWERS.

GERMANY. AUSTRIA-HUNGARY. TURKEY.

1. GERMANY.

GERMANY.

DREADNOUGHTS.

		Completed	Displacement (tons)	Speed (knots)	Armament	Complement
New Dreadnoughts.	Ersatz Kaiser Friedrich III .	(due) 1917	28,000	23	Guns: 8—15 inch; 16—6 inch; 12 smaller. Torpedo tubes (22 inch).	—
	Ersatz Worth .	1916				
	"T" . .					
1914. *Koenig class.*	Ersatz Brandenburg	1915	25,500	21.5	Guns: 10—12 inch, 45 cal.; 14—6 in.; 12—24 pdrs; 4—14 pdrs. (anti-aerial). Torpedo tubes (20 inch): 5.	1100
	Markgraf . .	1914				
	Koenig . .					
	Grosser Kurfürst.					
1912-13. *Kaiser class.*	Kaiser . .	1912	24,700	20	Guns: 12—12 inch, 50 cal.; 14—6 in.; 12—24 pdrs.; 4—14 pdrs. (anti-aerial). Torpedo tubes (20 inch): 5.	1088
	Friedrich Der Grosse					
	Kaiserin . .					
	Prinz Regent Luitpold .	1913				
	Koenig Albert .					
1911-12. *Helgoland class.*	Thüringen . .	1911	21,000 (normal)	20.5	Guns: 12—12 inch, 50 cal.; 14—6 in., 45 cal.; 14—24 pdrs. Torpedo tubes (20 inch): 6.	1106
	Helgoland . .					
	Ostfriesland .					
	Oldenburg . .	1912				

GERMANY.

DREADNOUGHTS.

		Completed	Displacement (tons)	Speed (knots)	Armament	Complement
1909-10. *Nassau class.*	Westfalen Nassau Posen Rheinland	1909 1910	18,900	19.5	Guns: 12—11 inch, 45 cal.; 12—6 inch, 45 cal.; 16—24 pdrs. Torpedo tubes (20 inch): 6.	963

BATTLE-CRUISERS.

			Completed	Displacement (tons)	Speed (knots)	Armament	Complement
Dread-noughts.		Derflinger Lützow Ersatz Hertha	1914 1915 1916	28,000	—	Guns: 8—12 inch, 50 cal.; 12—6 inch, 45 cal.; 12—24 pdrs. Torpedo tubes (22 inch): 5.	—
	1913.	Seydlitz	1913	25,000	26.5	Guns: 10—11 inch, 50 cal.; 12—6 inch; 12—24 pdrs.; 4—14 pdrs. (Anti-aerial). Torpedo tubes (20 inch): 4.	1108
	1911-12 *Moltke class.*	Moltke Goeben	1911 1912	23,000	27	Guns: 10—11 inch, 50 cal.; 12—6 inch; 12—24 pdrs. Torpedo tubes (20 inch): 4.	1107
	1910.	Von Der Tann	1910	19,400	25 28.1	Guns: 8—11 inch, 45 cal.; 10—6 inch; 16—24 pdrs. Torpedo tubes (18 inch): 4.	910

GERMANY.

BATTLESHIPS (PRE-DREADNOUGHTS).

		Completed	Displacement (tons)	Speed (knots)	Armament	Complement
Deutschland class.	Deutschland	1906	13,200	18	Guns: 4—11 inch, 40 cal.; 14—6.7 inch, 40 cal.; 20—24 pdrs.; 4—1 pdrs.; 4 Mach. Torpedo tubes (18 inch): 6.	729
	Hannover	1907				
	Pommern					
	Schleswig-Holstein	1908				
	Schlesien					
Braunschweig class.	Braunschweig	1904	13,200	18	Guns: 4—11 inch, 40 cal.; 14—6.7 inch, 40 cal.; 20—1 pdrs.; 8 Mach. Torpedo tubes (18 inch): 6.	691
	Elsass					
	Hessen	1905				
	Preussen					
	Lothringen	1906				
Wittelsbach class.	Wittlesbach	1902	11,830	18	Guns: 4—9.4 inch, 40 cal.; 18—6 in., 40 cal.; 12—15½ pdrs.; 12—1 pdrs. (revolver); 8 M. Torpedo tubes (18 inch): 6.	650 (as flagship 715).
	Wettin					
	Zahringen					
	Mecklenburg	1903				
	Schwaben					
Kaiser Friedrich class. (Refitted in 1907-8.)	Kaiser Friedrich III	1898	10,790	18	Guns: 4—9.4 inch, 40 cal.; 14—6 in., 40 cal. (*K. Karl* 18) 12—15½ pdrs.; 12—1 pdrs.; 8 Mach. Torpedo tubes (18 inch): 6.	658
	Kaiser Wilhelm II	1900				
	Kaiser Wilhelm Der Grosse					
	Kaiser Barbarossa	1901				
	Kaiser Karl Der Grosse					

GERMANY.

OLD BATTLESHIPS.

			Completed	Displacement (tons)	Speed (knots)	Armament	Complement
Old German Battleships.		Brandenburg	1891	10,060	17	Guns: 4—11 inch, 40 cal.; 2—11 in., 35 cal.; 8—4.1 inch, 30 cal.; 8—15½ pdrs.; 12—1 pdrs.; 4 Mach. Torpedo tubes (18 inch): 3.	552
		Worth	1892				
	Employed on Coast Defence in the Baltic.	Odin	1894	4,150	15	Guns: 3—old 9.4 inch, 35 cal.; 10—15½ pdrs.; 6—1 pdrs.; 4 Machine. Torpedo tubes (18 inch): 3.	
		Ægir	1895				
		Siegfried	1889	4,100			
		Beowulf	1890				
		Frithjof	1891				
		Heimdall	1892				
		Hildebrand					
		Hagen	1893				

GERMANY.

ARMOURED CRUISERS.

		Completed	Displacement (tons)	Speed (knots)	Armament	Complement
Roon class.	Roon . .	1905	9,050	21	Guns: 4—8.2 inch, 40 cal.; 10—6 inch, 40 cal.; 14—24 pdrs. 4 Machine. Torpedo tubes (18 inch): 4.	557
Prinz Adalbert class.	Prinz Adalbert	1903	9,050	21	Guns: 4—8.2 inch, 40 cal; 10—6 inch, 40 cal.; 10—15½ pdrs.; 10—1 pdrs.; 4 Maxims. Torpedo tubes (18 inch): 4.	557
	Prinz Heinrich	1900	8,930	20	Guns: 2—9.4 inch, 40 cal.; 10—6 inch, 40 cal.; 12—15½ pdrs.; 10—1 pdrs. Torpedo tubes (18 inch): 4.	501
	Fürst Bismarck	1897	10,700	19	Guns: 4—9.4 inch, 40 cal.; 12—6 inch, 40 cal.; 10—15½ pdrs.; 14—1 pdrs. Torpedo tubes: various; 3 sizes fitted for training purposes.	

GERMANY.

BIG PROTECTED CRUISERS.

		Completed	Displacement (tons)	Speed (knots)	Armament	Complement
1907-10 German Big Protected Cruisers.	Freya . .	1898	5,660	19	Guns: 2—8.2 inch, 40 cal; 6—6 inch, 40 cal.; 12—15½ pdrs.; 10—1 pdrs.; 4 Machine. Torpedo tubes (18 inch): 3.	465
	Hertha . .					
	Victoria Luise					
	Hansa . .		5,885			
	Vineta . .					
	Kaiserin Augusta	1892	6,060	20.7	Guns: 12—6 inch, 35 cal.; 8—15½ pdrs.; 8 Machine. Torpedo tubes: 5.	436

SMALL CRUISERS.

		Completed	Displacement (tons)	Speed (knots)	Armament	Complement
	Rostock .	1913				
1913-4 Small German Cruisers	Regensberg .	1914	4,900	28	Guns: 12—4.1 inch, 45 cal. Torpedo tubes (20 inch): 2.	373
	Graudenz .					
	Ersatz Gefion	1915				
	Ersatz Hela .					
	Ersatz Gazelle	1915				
	Ersatz Niobe					
1912 Small German Cruisers. *Breslau class.*	Breslau .	1912	4,550	28	Guns: 12—4.1 inch, 45 cal. Torpedo tubes (20 inch): 2.	379
	Stralsund .					
	Strassburg .					

GERMANY.

SMALL CRUISERS.

		Completed	Displacement (tons)	Speed (knots)	Armament	Complement
1909-10. *Kolberg class.*	Kolberg . Augsburg .	1909 1910	4,350	25.5	Guns: 12—4.1 inch, 40 cal.; 4—5 pdrs. 55 cal.; 4 Mach. Torpedo tubes (18 inch): 2.	362
1908. *Stettin class.*	Stuttgart . Stettin . .	1908	3,450	23.5	Guns: 10—4.1 inch, 40 cal.; 8—5 pdrs. 55 cal.; 4 Mach. Torpedo tubes (18 inch): 2.	295
1906-7. *Later Bremen class*	Danzig . .	1907	3,250	23	Guns: 10—4.1 inch, 40 cal.; 10–1 pdrs.; 4 Machine. Torpedo tubes (18 inch): 2.	286

365

GERMANY.

SMALL CRUISERS.

		Completed	Displacement (tons)	Speed (knots)	Armament	Complement
1904-5 Small German Cruisers.	Bremen . Hamburg Berlin . Luebeck Muenchen	1904 1905	3,250	23	Guns: 16—4.1 inch; 10—1 pdrs.; 4 Machine. Torpedo tubes (18 inch): 2.	286
1903 Small German Cruisers.	Frauenlob Arcona . Undine .	1903	2,715	21.5	Guns: 10—4.1 inch, 40 cal.; 14—1 pdrs.; 4 Machine. Torpedo tubes (18 inch): 2.	264
1901 Small German Cruisers.	Nymphe Thetis Medusa Amazone Ariadne	1901	2,660	21	Guns: 10—4.1 inch, 40 cal.; 14—1 pdr.; 4 Machine. Torpedo tubes (18 inch): 2.	247
1898-9 Small German Cruisers *Gazelle class.*	Gazelle . Niobe .	1898 1899	2,645	21	Guns: 10—4.1 inch, 40 cal; 14—1 pdr; 4 Machine. Torpedo tubes (18 inch): 2 submerged; *Gazelle* has 3.	264
Small German Protected Cruiser.	Gefion .	1894 (refitted 1905)	3,770	20	Guns: 10—4.1 inch; 6—4 pdrs; 8 machine. Torpedo tubes: 2.	302

GERMANY.

TORPEDO BOAT DESTROYERS.

1912-13 DESTROYERS.

S 36—31
V 30—25 } 820 tons. Speed: 32.5 kts.
S 24—13

Guns: 2—3.5 inch (24 pdrs.); 8—20 inch tubes (twins).

1912 DESTROYERS.

G 12—7
V 6—1 } 650 tons. Speed: 32.5 kts. (about).

Guns: 2—3.5 inch (24 pdrs.); 2 Machine. 5 tubes (20 inch). 2—24 pdrs. Torpedo tubes (20 inch): 5.

1911 DESTROYERS

G 197—192
G 175—174 } 654 tons. Speed: 31.5 kts.

Guns: 2—24 pdrs.; 2 Machine. Torpedo tubes (18 inch): 4.

S 179—176. 636 tons. Speed: 32 kts.

Guns: 2—24 pdrs.; 2 Machine. Torpedo tubes (18 inch): 4.

1911 DESTROYERS

V 191—186
V 185—180 } 350 tons. Speed: 32.5 kts.

Guns: 2—24 pdrs., 35 cal.; 2 Machine. Torpedo tubes (18 inch): 3.

1909 DESTROYERS.

S 168—165. 615 tons. Speed: 32 kts.

Guns: 2—24 pdrs.; 2 Machine. Torpedo tubes (18 inch): 3.

G 173—169. 620 tons. Speed: 32 kts.

Guns: 2—24 pdrs.; 2 Machine. Torpedo tubes (18 inch): 3.

V 164—162. 613 tons Speed: 32 kts.

Guns: 2—24 pdrs., 30 cal.; 2 Machine Torpedo tubes (18 inch): 3.

1908 DESTROYERS.

V 161—150. 680 tons. Speed: 30 kts.

Guns: 2—24 pdrs.; 2 Machine. Torpedo tubes (18 inch): 3.

GERMANY.

TORPEDO BOAT DESTROYERS.

1907 DESTROYERS.

S 149—138. 525 tons. Speed : 30 kts.

Guns : 1—24 pdr. ; 3—4 pdr. ; 2 Machine. Torpedo tubes (18 inch) : 3.

G 137. 572 tons. Speed : 32 kts.

Guns : 4—4 pdrs., 55 cal. ; 2 Machine. Torpedo tubes (18 inch) : 3.

G 136—132. 486½ tons. Speed : 28½ kts.

Guns : 4—4 pdrs., 55 cal. G 135 has special armament. 1—15 pdr. ; 2—4 pdrs.

1906 DESTROYERS.

S 131—114. 487 tons. Speed : 28 kts.

Guns : 3—4 pdrs. ; 2 Machine. Torpedo tubes (18 inch) : 3.

1902-1903 DESTROYERS.

G 113—108. 400 tons. Speed : 28 kts.

Guns : 3—4 pdrs. ; 2 Machine. Torpedo tubes (18 inch) : 3.

S 107—91. 400 tons. Speed : 26½ kts.

1894, etc., DESTROYERS.

D 9. 380 tons. Speed : 24½ kts.
Guns : 3—4 pdrs. Torpedo tubes (18 inch): 3.

D 8—7. 350 tons. Speed : 23 kts.
Guns : 3—4 pdrs. Torpedo tubes (18 inch) : 3

D 6—5. 320 tons. Speed : 23 kts.
Guns : 3—4 pdrs. Torpedo tubes (18 inch) : 3

D 4—3. 300 tons. Speed : 21 kts.
Guns : 3—4 pdrs. Torpedo tubes (18 inch) : 3

(A number of new boats believed to be building. Some boats building for foreign powers were also taken over),

GERMANY.

TORPEDO BOATS.

	tons.	kts.
G 88, 89	155	25
S 87, 82	155	25
S 81, 78, 76, 74	153	25
S 75, 78	145	26
S 73, 70, 68, 66	170	22
S 65, 58	153	22
T 57—49	130	22
T 47,—43 S 42	145	22
S 32	—	19

Guns : 1—4 pdr.
Torpedo tubes :
14 inch.

GERMANY.

SUBMARINES.

U 30—25. 800 tons.

Guns : 2—14 pdrs. (disappearing) ; 2—1 pdrs., high angle. Torpedo tubes : **4.**

U. 24—17 $\frac{650}{750}$ tons.

Guns : 1—14 pdr. (disappearing) ; 2—1 pdrs., high angle. Torpedo tubes : 4.

U 16—13. $\frac{250}{300}$ tons.

Guns : 2—1 pdrs., high angle. Torpedo tubes (18 inch) : 3.

U 12—8. $\frac{250}{300}$ tons.

Guns : 2—1 pdrs., high angle. Torpedo tubes (18 inch) : 3.

U 7—2. $\frac{210}{250}$ tons.

Torpedo tubes : (18 inch) : 2.

U 1. $\frac{197}{236}$ tons.

Torpedo tubes (18 inch) : 1.

*Of these U 29 ; also 18, 15, 9, are admitted as sunk. About twelve others are also believed to have been destroyed. The number of new boats in hand or completing is understood to be about 60.

The whole of the above are fitted with submarine signalling apparatus. Average time to submerge is 25 to 30 seconds. Some boats carry searchlights. Early boats have 6-cyl. Nuremburg motors. Later have Krupp and Nuremburg motors ; but some are fitted with Augsburg 2-cycle motors (M.A.N.) All boats after and including *U* 13 have sleeping accommodation. The programme provides for 72 boats by end of 1917.

GERMANY.

MISCELLANEOUS.

		Completed	Displacement (tons)	Speed (knots)	Armament	Complement
Unprotected Cruisers.	Geier* . Condor . See-Adler	1892	1,630	15–17	Guns : 8—4.1 inch ; 5—1 pdrs. Torpedo tubes (14 inch) : 2.	165
Gunboats. . .	Eber . Panther	1903 1901	1,000	14	Guns : 2—4.1 inch, 40 cal. ; 6—1 pdrs. ; 2 Machine.	125
River Gunboat. .	Otter .	1909	260	14	Guns : 2—4 pdrs.	
Surveying Ship .	Planet .	1905	650	10	—	

* Interned.

GERMANY.

MISCELLANEOUS.

		Completed	Displacement (tons)	Speed (knots)	Armament	Complement
Special Mine Ships. .	Albatross		2,200	21	Guns: 8—3.5 inch, 24 pdrs., 35 cal.	189
	Nautilus		1,970	20	Guns: 8—3.5 inch, 35 cal.	189
Mining Service. .	Pelikan		2,360	15	Guns: 4—15½ pdrs.	
	Rhein .		400	10	Nil.	
Subsidiary Service. .	Schwalbe		1,120	13—14	Guns: 8—4.1 inch; 5—1 pdrs. Torpedo Tubes (14 inch): 2.	
	Blitz . / Pfeil .		1,390	15	Guns: 6—15 pdr.	
Special Service Tenders.	Drache .	1908	920	15	Guns: 8—4.1 inch.	
	Hai .	1907	640	12	Guns: 8—4 pdrs.	
	Delphin	1906	450	11	Guns: 4—24 pdrs.	
	Fuchs .	1905	640	12	Guns: 2—4.1 inch.	
Special Submarine Ship.	Vulkan	1908	—	11	Nil.	

Plate XCIII

GERMAN "BRAUNSCHWEIG" (Battleship).

Plate XCIV

GERMAN "GROSSER KURFÜRST" (Battleship).

GERMAN "KAISER" (Battleship)

GERMAN "WETTIN" (Battleship).

GERMAN "KAISER BARBAROSSA" (Battleship).

Plate XCVIII

GERMAN "SEYDLITZ" Battle Cruiser.

Plate XCIX

GERMAN "VON DER TANN" (Battle Cruiser). *Cribb*

GERMAN "DERFFLINGER" (Battle Cruiser).

Plate CI

GERMAN "STETTIN" (Light Cruiser). *Cribb*

2. AUSTRIA-HUNGARY.

AUSTRIA-HUNGARY

BATTLESHIPS.

Class	Ship	Completed	Displacement (tons)	Speed (knots)	Armament	Complement
Viribus Unitis class.	Szent Istvan .	1915	20,000	21	Guns: 12—12 inch, 45 cal; 12—6 inch; 18—12 pdrs.; 2—12 pdrs. (Anti-aerial). Torpedo tubes (21 inch): 4.	1000
	Prinz Eugen .	1914				
	Tegetthof .	1913				
	Viribus Unitis .	1912				
Radetzky class.	Radetzky . .	1911	14,500	20	Guns: 4—12 inch, 45 cal; 8—9.4 inch, 45 cal.; 20—4.1 inch, 50 cal.; 6—12 pdrs.; 2—12 pdrs., 18 cal. (Anti-aerial). Torpedo tubes (18 inch): 3.	880
	Zrinyi . .					
	Erzherzog Franz Ferdinand .	1910				
Erzherzog Karl class.	Erzherzog Ferdinand Max.	1907	10,600	20.25	Guns: 4—9.4 inch, 40 cal.; 12—7.6 inch, 42 cal.; 12—12 pdrs., 45 cal.; 2—12 pdrs., 18 cwt. (Anti-aerial). Torpedo tubes (18 inch): 2.	700
	Erzherzog Friedrich	1906				
	Erzherzog Karl	1905				
Habsburg class	Habsburg .	1903	8,340	19	Guns: 3—9.4 inch, 40 cal.; 12—6 inch, 40 cal.; 10—12 pdrs.; 2—12 pdrs. 18 cal. Torpedo tubes (18 inch): 2.	
	Arpad . .					
	Babenberg .	1904				
	Budapest .	1897	5,600	17	Guns: 4—9.4 inch, 40 cal.; 6—6 inch, 40 cal.; 2—12 pdrs. 18 cwt.; 14—3 pdrs. Torpedo tubes (18 inch): 2.	400
	Wien . .	1896				
	Monarch . .					

AUSTRIA-HUNGARY.

CRUISERS.

		Completed	Displacement (tons)	Speed (knots)	Armament	Complement
Armoured Cruisers.	Sankt Georg	1906	7,400	21	Guns: 2—9.4 inch, 40 cal.; 5—7.6 inch, 42 cal.; 4—6 inch, 40 cal.; 9—12 pdrs.; 2—12 pdrs. (Anti-aerial); 10—3 pdrs. Torpedo tubes (18 inch): 2.	600
	Kaiser Karl VI	1899	6,325	20	Guns: 2—9.4 inch, 40 cal.; 8—6 inch, 40 cal.; 2—12 pdrs. 18 cwt., Anti-aerial); 16—3 pdrs.; 1 Mach. 2—12 pdrs. (Boat guns). Torpedo tubes (18 inch): 2.	450
Kaiserin und Königin Maria Theresia		Reconstructed 1907	5,200	17	Guns: 2—7.6 inch, 40 cal.; 8—6 inch, 35 cal; 16—3 pdrs.; 1 Mach. Torpedo tubes (18 inch): 4.	450
Kaiser Franz Josef . .		Reconstructed 1905–6	4,060	18	Guns: 2—6 inch, 40 cal.; 6—6 inch, 35 cal.; 16—3 pdrs.; 1 Machine. Torpedo tubes (14 inch): 4.	450

AUSTRIA-HUNGARY.
CRUISERS.

		Completed	Displacement (tons)	Speed (knots)	Armament	Complement
1911. *Improved Spaun class.*	Saida . Helgoland Novara	1914	3,500	27	Guns: 9—4.1 inch; 2—12 pdrs. 18cwt. Torpedo tubes (18 inch): 4.	320
	Admiral Spaun	1910	3,500	26	Guns: 7—4.1 inch. Torpedo tubes (18 inch): 2.	320
Zenta class. . .	Szigetvar Aspern .	1901 1900	2,350 2,400	20	Guns: 8—4.7 inch, 40 cal.; 10—3 pdr. Torpedo tubes (18 inch): 2.	300

AUSTRIA-HUNGARY.
TORPEDO GUNBOATS.

		Completed	Displacement (tons)	Speed (knots)	Armament	Complement
	Planet Trabant		480 540	20	Guns: 2—12 pdrs.; 8—3 pdrs. Torpedo tubes: 3.	78
	Satellit .		540	22	Guns: 1—12 pdr.; 8—3 pdrs. Torpedo tubes: 3.	61
	Meteor .		350	21	Guns: 8—3 pdrs. Torpedo tubes: 4.	49
	Magnet		510	26	Guns: 6—3 pdrs. Torpedo tubes: 3.	80

AUSTRIA-HUNGARY.

18 Destroyers.

No.	Type	Date	Displacement	H.P.	*Max.* speed	Complement
			tons		kts.	
6	*Tatra*	'11-'13	800	17,000	32 (*t*)	...
6	*Huszar* (Y)	'07-'10	390	6000	28	64
2		'07-'09				
3		'06-'08				
1		'05-'07				

(Y) = Yarrow type.

Tatra, Balaton, Csepel, Lika, Orjen, Triglav.

Huszar class: named ***Huszar, Ulan, Streiter, Wildfang, Uskoke, Scharfschütze,*** also ***Csikos, Dinara, Pandur, Reka, Turul, Velebit.***

11 Submarines.

No. in Class	Type	Completed	Displacement. Surface	Surface H.P.	*Max.* speed above and below	Tubes	Complement
			tons		kts.		
5	U7 – U 11 (G)	'14	.../520	...	...	...	...
2	U 6, U 5 (H)	'10	236/273	500	11½–10	2	16
2	U 4, U 3 (G)	'10	237/300	600	11½–8	2	17
2	U 2, U 1 (L)	'10	220/250	720	12–7½	3	...

(H) = Holland type. (G) = Germania, Krupp type. (L) = Lake type.

51 Sea-going, 40 Coastal (= total of 91) Torpedo Boats.

Type	Date	Displacement.	I.H.P.	*Max.* speed.	Complement.	Tubes.
Sea-going –						
27 new (74 T–81 T, 82 F–97 F, 98 M–100 M)	1911-13	250	5000	28(*t*)	...	2
24 *Kaiman* (*Yarrow*)	1905-'08	200	3000	26	29	2
Coastal –						
12 *No. 1–12* (oil fuel)	1909-10	110	2600	27	20	2
1 *No. 18* (*Yarrow*)	1896	107	1800	26	20	2
4 *No. 13–17* (*Yarrow*)	'98-'99	115	1800	26	20	3
1 *No. 45* (*Schichau*)	1896	134	1800	26	20	3
23 *No. 19–40* (*Schichau*)	'92-'86	78	900	19	16	2

AUSTRIA-HUNGARY.

MISCELLANEOUS.

		Displacement (tons)	Speed (knots)	Armament
Subsidiary Local Defence . .	K. E. Rudolf	6.940	16	Guns: 3—old 12 in.; 6—4.7 inch.
	Mars . .	7,390	16	Guns: 6—old 9.4 inch Krupp; 5—old 6 inch.
Torpedo Depot Ship . . .	Gaea . .	12,000	18	Guns: 4—4.7 inch; 4—12 pdrs.
Submarine Depot Ships . . .	Pelikan .	2,440	14	—
	Gigant .	260	11	—
Mining Ships.	Basilisk .	310	11	Guns: 2—3 pdrs.
	Salamander .	268	10	Guns: 2—3 pdrs.
	Dromedar .	175	10	—
Repair Ship	Vulkan .	950	15	Nil.

AUSTRIAN "VIRIBUS UNITIS" (Battleship).

PLATE CIII

AUSTRIAN "RADETZKY" (Battleship). *Cribb*

3. TURKEY.

TURKEY.

BATTLESHIPS.

		Completed	Displacement (tons)	Speed (knots)	Armament	Complement
[illegible]	Mesudiyeh		9 1[illegible]	[illegible]	2-9.2" 12-6" 14-3" 10-6pdr 2-3pdr 2-1pdr	600
Old Battleships.	Häirredin Barbarosse (ex *Kürfurst Friedrich Wilhelm*) Torgud Reis . (ex *Weissenburg*		10,060	17	Guns: 4—11 inch, 40 cal.; 2—11 in., 35 cal.; 8—4.1 in. 35 cal.; 8—15½ pdrs.; 4 Machine. Torpedo tubes (18 inch): 2.	
Coast Defence Battleship.	Muin-i-Zaffer	Rebuilt 1904–7	2,400	12	Guns: 4—6 inch, 40 cal.; 6—3 in.; 14½ pdr.; 10—6 inch; 2—1 pdrs.	220
CDB	[illegible]		[illegible]	13	3-5.9" 7-4.7" [illegible]	300

TURKEY.

BATTLESHIPS.

		Completed	Displacement (tons)	Speed (knots)	Armament	Complement
Protected Cruisers	Hamidieh . .	1904	3,800	22	Guns: 2—6 inch, 45 cal.; 8—4.7 in., 50 cal.; 2—3 pdrs.; 2—1 pdrs. Torpedo tubes (18 inch): 3.	302
Torpedo Gunboats.	Peik-i-Shevket Berki-Satvet .	1907	775	22	Guns: 2—4.1 inch; 6—6 pdrs.; 2—1 pdrs.; 5 Machine. Torpedo tubes (18 inch): 3.	—
	Pelenk-i-Deria	1891	775	19	Guns: 2—4 inch; 6—3½ pdrs. Torpedo tubes (14 inch): 3.	
Gunboats.	Marmaris . .	1907	500	14	Guns: 4—9 pdrs.; 2—1 pdrs.	
	Nerchehir; Tachkeupru	1909	213	12	Guns: 2—3 pdrs.	
	Zuhof . . .		650	12	Guns: 4—1 pdrs. Torpedo tubes: 2.	
	Castamoni; Yozgad.	1906	185	12	Guns: 2—3 pdrs.	
	Nourilbahir . .	1903	198	12	Guns: 4—12 pdrs.; 2—3 pdrs. Tubes 2	

Also several other Gunboats: 1 Torpedo Depot Ship: Ali Pacha; 5 Transports; 3 Despatch Vessels: Izzedin, Ismail, Thalia; 1 Mine Ship: Muzrat; 1 Repair Ship: Tiri-Mujgian.

TURKEY.

10 Destroyers + 12 *pro.*

No.	Type	Date	Displacement	*Max.* speed	Complement
			tons	kts.	
12		*pro.*			
4	*Yadikiar Milet* (S)	'09-'10	620	35	83
4	*Samsoun* (L)	'07-'10	305	28	

(S)=Schichau. (L)=Chantiers de la Loire.

4 *Schichau* boats. ***Yadikar-Milet, Mouavenet-Millieh, Noumounel-Hamiet,*** and ***Gaïret-Vatandeh.***
Samsoun, Basra, Tassos, Yar-Hissar.

9 Torpedo Boats.

No.	Type	Date	Displacement	H.P.	*Max.* speed	T tubes
			tons		kts.	
4	*Hamid Abad* (C)	'06-'07	97	1900	26	3
6	*Antalia* (A)	'04-'07	165	2200	27	1
1	*Younis* (A)	1901	145	1900	27	1

(A)=Ansaldo. (C)=Creusot. (Five other boats lost in Italian War, and one in the Greek War.)

Hamid-Abad, Sultan-Hissar, Sivre-Hissar and ***Timur-Hissar.***
Qutahia, Mosoul, Tocad and ***Ak-Hissar.***
The 145 ton Ansaldo boat is named ***Younis.***

TURKISH "HAMIDIEH" (Cruiser).

Hopkins

III. THE NEUTRAL POWERS.

AMERICA, AND THE OTHERS IN ALPHABETICAL ORDER.

1. AMERICA.

UNITED STATES.

DREADNOUGHTS.

		Completed	Displacement (tons)	Speed (knots)	Armament	Complement
1913 Dreadnoughts.	Idaho . Mississipi . California .	Due 1917				
Pennsylvania class.	Pennsylvania Arizona . .	Due 1916	— 31,400	— 21	Guns: 12—14 inch, 45 cal.; 22—5 inch, 51 cal.; 4—3 pdrs. Torpedo tubes (12 inch): 4.	
1912 Dreadnoughts.	Oklahoma . Nevada .	1915	27,500	20.5	Guns: 10—14 inch, 45 cal.; 21—5 inch, 51 cal.; 4—3 pdrs. (saluting); 2—1 pdr. Torpedo tubes (21 inch): 4.	863
1911 Dreadnoughts. *Texas class.*	Texas . New York	1913 1914	27,000	21	Guns: 10—14 inch, 45 cal.; 21—5 inch, 51 cal.; 4—3 pdrs. (saluting); 2—1 pdr. Torpedo tubes (21 inch): 4.	1,015
1910 Dreadnoughts. *Arkansas class.*	Arkansas . Wyoming .	1912	26,000	20.5	Guns: 12—12 inch, 50 cal.; 21—5 inch, 51 cal.; 4—3 pdrs. (saluting); Torpedo tubes (21 inch): 2.	1,036
1909 Dreadnoughts. *Utah class.*	Utah. . Florida .	1911	21,825	20.75	Guns: 10—12 inch, 45 cal.; 16—5 inch, 51 cal.; 4—6 pdrs.; 2—1 pdrs.; 2 Machine; 2—3 inch Ldg. guns. Torpedo tubes (21 inch): 2.	940

UNITED STATES.

DREADNOUGHTS.

		Completed	Displacement (tons)	Speed (knots)	Armament	Complement
1907 Dreadnoughts. *Delaware class.*	Delaware . North Dakota	1910	20,000	21	Guns: 10—12 inch, 45 cal.; 14—5 inch, 50 cal. R. F.; 4—3 pdrs. (saluting). Torpedo tubes (21 inch): 2.	927
1906 Dreadnoughts. *South Carolina class.*	South Carolina Michigan .	1909	16,000	18.5	Guns: 8—12 inch, 45 cal.; 22—3 inch, 14 pdrs.; 2—3 pdrs. S.A. (saluting). Torpedo tubes (21 inch): 2.	805
BATTLESHIPS.						
1904 Battleships. (*First Line*). *Kansas class.*	New Hampshire Kansas . Vermont . Minnesota .	1908 1907 1906	16,000	18	Guns: 4—12 inch, 45 cal.; 8—8 inch, 45 cal.; 12—7 inch, 50 cal.; 20—3 inch, 14 pdrs.; 4—6 pdrs. S.A. Torpedo tubes (21 inch): 4.	953 (Flagship 1,013.
1903 Battleships. *Second Line.*) *Louisiana class.*	Louisiana . Connecticut	1906	16,000	18	Guns: 4—12 inch, 45 cal.; 8—8 inch, 45 cal.; 12—7 inch. 50 cal.; 20—3 inch, 14 pdrs.; 4—6 pdrs. (saluting). Torpedo tubes (21 inch): 4.	953 (Flagship 1,008).

UNITED STATES.

BATTLESHIPS.

		Completed	Displacement (tons)	Speed (knots)	Armament	Complement
1902 Battleships. (*Second Line.*) *New Jersey class.*	New Jersey Georgia . Rhode Island Virginia . Nebraska .	1906 1907	14,948	19	Guns: 4—12 inch, 40 cal.; 8—8 inch, 45 cal.; 12—6 inch, 50 cal.; 12—3 inch, 14 pdrs.; 4—6 pdrs. (saluting). Torpedo tubes (21 inch): 4.	910
1899 Battleships. (*Second Line.*) *Maine class.*	Maine . Missouri . Ohio .	1902 1903 1904	12,500	18	Guns: 4—12 inch, 40 cal.; 16—6 inch, 50 cal; 6—3 inch, 14 pdrs.; 4—6 pdrs. Torpdeo tubes (18 inch): 2.	798
1896 Battleships. (*Second Line.*)	Alabama Illinois . Wisconsin	1906 1901	11,552	16	Guns: 4—13 inch, 35 cal.; 14—6 inch, 40 cal.; 4—3 inch, 50 cal.; 4—6 pdrs. (saluting). No torpedo tubes.	693
Kearsage class.	Kearsage Kentucky	1900	11,520	16	Guns: 4—13 inch; 4—8 inch, 35 cal.; 18—5 inch, 40 cal.; 4—6 pdrs. Torpedo tubes: (1—18 inch in *Kearsage*).	726
	Iowa .	1897	11,346	17	Guns: 4—12 inch, 35 cal.; 8—8 inch, 35 cal.; 10—4 inch; 4—6 pdrs. No tubes.	683
Old Battleships. (*second Line.*) *Indiana class.*	Indiana . Massachusetts Oregon .	1895 1895 1896	10,288	17	Guns: 4—13 inch; 8—8 inch, 35 cal.; 12—3 inch, 14 pdrs., 50 cal.; 4—6 pdrs. No torpedo tubes.	178

UNITED STATES.

COAST DEFENCE VESSELS.

		Completed	Displacement (tons)	Speed (knots)	Armament	Complement
Ozark class.	Ozark (ex *Arkansas*) Tonopah (ex *Nevada*) Talahassee (ex *Florida*) Cheyenne (ex *Wyoming*)		3,225		Guns: 2—12 inch, 40 cal.; 4—4 inch, 50 cal.; 3—6 pdrs. No torpedo tubes.	
	Monterey . .	1893	4,084	14	Guns: 2—12 inch, 35 cal.; 2—10 in., 30 cal.; 6—6 pdrs.	
	Monadnock . .	1885	3,990	14.5	Guns: 4—10 inch. 2—4 inch; 4—6 pdrs.	
	Miantonomoh . . Terror . .	—	3,990	12	Guns: 4—10 inch; 2—6 pdrs.	
	Puritan					230

UNITED STATES.

CRUISERS.

		Completed	Displacement (tons)	Speed (knots)	Armament	Complement
1903 Armoured Cruisers. *Washington class.*	North Carolina Montana . Washington Tennessee .	1908 1907	14,500	22	Guns: 4—10 inch, 40 cal.; 16—6 in., 50 cal.; 24—3 in., 14 pdrs.; 4—6 pdrs. Torpedo tubes (21 inch): 4.	945 (flagship 990)
1901 Armoured Cruisers. *California class.*	West Virginia Colorado . Pittsburg . Maryland . California . South Dakota	1905 1907 1908	13,680	22	Guns: 4—8 inch, 45 cal.; 14—6 in., 50 cal.; 18—3 in., 50 cal.; 4—3 pdrs. Torpedo tubes (18 inch): 2.	878 (flagship 921
1902 First Class Cruisers *St. Louis class.*	Charleston . Milwaukee . St. Louis .	1905 1906	9,700	21.5	Guns: 14—6 inch, 50 cal.; 18—3 in., 14 pdr.: 4—3 pdr. No torpedo tubes.	727
First Class Cruisers.	Brooklyn .	1896 (Reconstructed 1907)	9,215	21	Guns: 8—8 inch, 35 cal.; 12—5 in., 40 cal.; 4—6 pdrs.	572
	Saratoga .	1893 (Reconstructed 1907)	8,150	21	Guns: 4—8 inch, 45 cal.; 10—5 in., 50 cal.; 8—3 inch, 14 pdrs.; 4—3 pdrs.	525

UNITED STATES.

CRUISERS.

		Completed	Displacement (tons)	Speed (knots)	Armament	Complement
Old Second Class Cruisers.	Olympia .	1895	5,870	21.7	Guns: 4—8 inch, 35 cal.: 10—5 in., 40 cal.: 4—6 pdrs.	446
	Baltimore .	—	4,413	20	Guns: 12—6 inch, 40 cal.; 4—6 pdrs.	374
	Columbia . Minneapolis	—	7,350	22.8	Guns: 3—6 inch, 45 cal.; 8—4 in., 40 cal.; 4—6 pdrs.	524
1905 Third Class Cruisers.	Salem . Birmingham Chester .	1908	3,750	24	Guns: 2—5 inch, 50 cal.; 6—3 in.; 2—6 pdrs. in *Birmingham* (3—2 pdrs. in others). Torpedo tubes (21 inch): 2.	373
1900 Third Class Cruisers.	Chattanooga Denver Galveston Cleveland Des Moines Tacoma	1904	3,200	16.5	Guns: 10—5 inch, 50 cal.; 8—6 pdrs. No torpedo tubes.	309
1895 Third Class Cruisers.	Albany . New Orleans	1900 1898	3,430	20.25	Guns: 10—5 inch, 50 cal.; 8—3 pdrs.	346
1889 Third Class Protected Cruisers.	Raleigh . Cincinnati .	1894	3,213	18	Guns: 11—5 inch, 40 cal.: 3—3 pdrs	

UNITED STATES.

GUNBOATS.

		Completed	Displacement (tons)	Speed (knots)	Armament	Complement
Gunboats	Paducah		1,085	12.9	Guns: 6—4 inch, 50 cal.; 4—6 pdrs.; 2—1 pdrs.	162
	Wheeling		1,000	13	Guns: 6—4 inch, 40 cal.; 4—3 pdrs.; 2—1 pdrs.	163
	Annapolis Princetown Vicksburg		1,000	12	Guns: 6—4 inch; 4—6 pdrs.	160
	Helena Wilmington		1,397	13	Guns: 8—4 inch, 40 cal.; 4—3 pdrs.	187
	Nashville		1,371	16	Guns: 8—4 inch; 4—6 pdrs.; 2—1 pdrs.; 2 Machine.	182
	Petrel		892	11.5	Guns: 4—4 inch; 2—3 pdrs.; 2—1 pdrs.	142
	Sacramento		1,700	16	Guns: 3—4 inch; 2—3 pdrs.; 2—1 pdrs.	

UNITED STATES.

TORPEDO BOAT DESTROYERS.

	Speed (knots)	Armament	Complement
6 *Cushing class* : McDougal, Nicholson, O'Brien, Winslow, Cushing, Ericsson	29	Guns : 4—4 inch, 50 cal. Torpedo tubes (18 inch) : 8.	98
8 *Alwyn class* . Cassin, Cummings, Alwyn, Balch, Benham, Parker, Duncan, Downes . .	29	Guns : 4—4 inch, 50 cal. Torpedo tubes (18 inch) : 8.	98
21 *Paulding* : Drayton, Jowett, Jenkins, Paulding, Trippe, Ammen, Burrows, Jarvis, McCall, Beale, Mayrant, Patterson, Warrington, Henley, Perkins, Sterrett, Walke, Fanning, Monaghan, Roe, Terry . . .	29½	Guns : 5—3 inch, 50 cal. Torpedo tubes (18 inch) : 6.	79
5 *Smith class* : Flusser, Reid, Lamson, Smith, Preston	28	Guns : 5—3 inch, 50 cal. Torpedo tubes (18 inch) : 3.	83
Old Destroyers. 9 *Bainbridge class* : Stewart, Bainbridge, Barry, Chauncey, Dale, Decatur, Paul Jones, Perry, Preble .	29	Guns : 2—3 inch, 5—6 pdrs. Torpedo tubes (18 inch) : 2.	72
Old Destroyers. 3 *Truxton class* : Truxton, Whipple, Worden	30	Guns : 2—3 inch ; 6—6 pdrs. Torpedo tubes (18 inch) : 2.	75
Old Destroyers. 2 *Hull class* : Hopkins, Hull . .	29	Guns : 2—3 inch ; 5—6 pdrs. Torpedo tubes (18 inch) : 2.	75
Old Destroyers. 2 *Lawrence class* : Lawrence, Macdonough	30	Guns : 7—6 pdrs. Torpedo tubes (18 inch) : 2.	75

UNITED STATES.

TORPEDO BOATS.

		Displacement (tons)	Speed (knots)	Armament	Complement
3 *Barney class*	Barney, Biddle, Bagley	167	28.5	Guns : 3—3 pdrs. Torpedo tubes (18 inch) : 3.	27
7 *Shubrick class*	Shubrick, Thornton, Stockton, De Long, Blakeley, Wilkes, Tingey . . .	165	26	Guns : 3—1 pdrs. Torpedo tubes (18 inch) : 3.	29
2 *Dahlgren class*	Craven, Dahlgren .	146	22.5	Guns : 4—1 pdrs. Torpedo tubes (18 inch) 2.	26
2 *Morris class*	Davis, Morris .	132	24	Guns : 3—1 pdrs. Torpedo tubes (18 inch) : 3.	24
1 Stringham		340	30	Guns : 4—6 pdrs. Torpedo tubes : (18 inch) 2.	55
1 Goldsborough		248	30	Guns : 4—6 pdrs. Torpedo tubes (18 inch) : 2.	61
1 Farragut		273	30	Guns : 4—6 pdrs. Torpedo tubes : (18 inch) 2.	62
1 Bailey		235	30	Guns : 4—6 pdrs. Torpedo tubes (18 inch) : 2.	57
1 Dupont		165	28	Guns : 4—1 pdrs. Torpedo tubes (18 inch) : 3.	30
1 Mackenzie		65	20	Torpedo tubes (18 inch) : 2.	16
1 Gwin		46	20	Torpedo tubes (18 inch) : 2.	9

UNITED STATES.

SUBMARINES.

	Displacement (tons)	H.P. Surface	Armament	Complement
4 *New* Boats	740	—	—	—
1 M 1 (H)	—	—	—	—
7 L 5—7 (L)	575	—	—	—
L 1—4 (H)	—	—	—	—
8 K 1—K 8 (H)	390 / 520	950	—	—
3 H 1—H 3 (H)	430	800	Torpedo tubes : 4.	—
1 G 4 (Li)	358 / 458	1000	Torpedo tubes : 4.	—
3 G 1—G 3 (L)	540	1200	Torpedo tubes : 6.	—
4 F 1—F 4 (H)	350 / 430	—	Torpedo tubes : 2.	—
2 E 1, E 2 (H)	350 / 430	—	Torpedo tubes : 2.	—
3 D 1—D 3 (H)	280 / 345	600	Torpedo tubes : 2.	15
5 C 1—C 5 (H)	238 / 275	500	Torpedo tubes : 2.	10
3 B 1—B 3 (H)	170	250	Torpedo tubes : 2.	10
7 A 1—A 7 (H)	120	160	Torpedo tubes : 1.	7

Some very large seagoing submarines are projected or ordered recently.

UNITED STATES.

MISCELLANEOUS.

		Displacement (tons)	Speed (knots)	Armament
Tenders to Destroyers	Melville	6,300	—	—
	Dixie	6,114	14.5	Guns: 10—3 inch; 2—6 pdrs.
	Pompey	3,085	10.5	Nil.
	Iris	6,100	10	Nil.
Tenders to Submarines	Bushnell	3,600	14	—
	Fulton	1,408	12.25	Guns: 4—3 inch R.F.
	Castine	1,177	16	Guns: 2—6 pdrs. R.F.
	Severn	1,175	—	Nil.
	Alert	1,100	10	Guns: 6—4 inch; 4—6 pdrs.
	Mohican	1,900	—	Guns: 4—6 pdrs. R.F.
Transports	General Alava	1,115	10.5	Guns: 2—6 pdrs.
	Buffalo	6,900	14.5	Guns: 2—5 inch; 4—4 inch; 6—3 pdrs.
	Prairie	6,620	14.5	Guns: 10—3 inch.
	Rainbow	4,360	12	Guns: 6—6 pdrs.; 6—1 pdrs.
	Hancock	8,500	—	Guns: 2—6 pdrs.
Hospital Ships	Relief	3,300	15	Nil.
	Solace	3,300	15	Nil.

UNITED STATES.

AUXILIARIES.

		Displacement (tons)	Speed (knots)	Armament
Fuel Ships	Jason, Orion, Nereus, Proteus, Neptune, Cyclops, Jupiter	Average 19,300	14	
	Kanawa, Maumee	14,500	14	
	Hector, Mars, Vulcan	11,230	12.5	
	Prometheus	12,585	16	
	~~And 12 Ordinary Colliers.~~			
Supply Ships	Celtic	8,000	10.5	Guns: 2—6 pdrs.
	Culgoa	6,100	12.5	Guns: 2—6 pdrs.
	Glacier	8,325	12.5	Guns: 1—3 pdr.
	Supply	4,325	9	Guns: 6—6 pdrs.; 4—1 pdrs.
(Surveying)	Eagle	434	15.5	Guns: 2—6 pdrs.
(Presidential Yacht)	Mayflower	2,690	16.8	Guns: 6—6 pdrs.
	Scorpion	775	17.75	Guns: 4—6 pdrs.
	Sylph	152	15	Nil.
	Yankton	975	14	Guns: 4—3 inch.
(Experimental Mine School Ship)	Restless	—	13	Guns: 1—1 pdr.

UNITED STATES.

AUXILIARIES.

			Displacement (tons)	Speed (knots)	Armament
Special Type.	Torpedo Exp. Ships. .	Dolphin . .	1,486	13.5	Guns : 6—6 pdrs.
		Lebanon . .	3,285	10	Guns : 2—6 pdrs.
		Montgomery	—	—	Guns : 2—6 pdrs. Torpedo Tubes · 2 submerged ; 2 above water
		Vesuvius . .	—	—	
		Vestal . .	12,565	16	None.
		Panther . .	3,380	13.5	Guns : 2—6 pdrs
		San Francisco .	4,098	19	Guns : 8—5 inch ; 2—6 pdrs.
Tugs . .	First Class Sea-going Tugs	Acushnet . .	769	—	Guns : 2—1 pdrs.
		Snohomish .	795	—	Guns : 2—3 pdrs.
	Ocean-going Tugs . .	Sonoma .	1,120	13	
		Ontario .			

PLATE CV

U.S.S. "TEXAS" (Battleship). *Müller*

U.S.S. "WYOMING" (Battleship). *Müller*

U.S.S. "FLORIDA" (Battleship).

Muller

Plate CVIII

U.S.S. "DELAWARE" (Battleship). *Müller*

PLATE CIX

U.S.S. "MICHIGAN" (Battleship) *Müller*

PLATE CXIII

U.S. SUBMARINE G 4 *Müller*

U.S.S. "MINNESOTA" (Battleship). *Müller*

U.S.S. "NEW JERSEY" (Battleship). *Müller*

U.S.S. "MONTANA" (Cruiser.) *Müller*

2. THE OTHER NEUTRAL POWERS IN ALPHABETICAL ORDER.

ARGENTINA

BRAZIL
BULGARIA

CHILI
CHINA
COLOMBIA
CUBA

DENMARK

ECUADOR
EGYPT

GREECE

HOLLAND

ITALY

MEXICO

NORWAY

PANAMA
PERSIA
PERU
PORTUGAL

ROUMANIA

SPAIN
SWEDEN

URUGUAY

VENEZUELA

ARGENTINA.

ARMOURED SHIPS.

	Completed	Displacement (tons)	Speed (knots)	Armament	Complement
Rividavia . . . . . Moreno	1911	28,000	22.5	Guns: 12—12 inch, 50 cal.; 12—6 inch; 16—4 inch. Torpedo tubes (21 inch): 2.	1,050
Pueyrredon . . .	1901	6,773	20	Guns: 2—10 inch, 40 cal.; 10—6 inch, 40 cal.; 6—4.7 inch, 40 cal.; 10—6 pdrs.; 12—1 pdrs.; 2 Max. Torpedo tubes (18 inch): 4.	500
General Belgrano . .	1899	7,069	20	Guns: 2—10 inch, 40 cal.; 14—6 inch, 40 cal.; 2—12 pdrs.; 10—6 pdrs.; 12—1 pdrs.; 2 Maxims. Torpedo tubes (18 inch): 4.	500
San Martin . . .	1898	6,773	18.5	Guns: 4—8 inch, 40 cal.; 10—6 inch, 40 cal.; 6—4.7 inch, 40 cal.; 12—6 pdrs.; 10—1 pdrs.; 2 Max. Torpedo tubes: 4.	500
Garibaldi	1896	6,732	18.5	Guns: 2—10 inch, 40 cal.; 10—6 inch, 40 cal.; 6—4.7 inch, 40 cal.; 10—6 pdrs.; 10—1 pdrs.; 2 Max. Torpedo tubes: 4.	500

ARGENTINA.

BATTLESHIPS AND CRUISERS.

		Completed	Displacement (tons)	Speed (knots)	Armament	Complement
Coast Defence Battleships.	Independencia / Libertad .	1893	2,336	13.2	Guns: 2—9.4 inch, 35 cal.; 4—4.7 inch, 40 cal.; 4—3 pdrs.; 2—1 pdrs. Torpedo tubes: 2.	225
	Almirante Brown .	Reconstructed 1897	4,267	14	Guns: 10—6 inch, 50 cal.; 4—4.7 inch, 50 cal.; 8—3 pdrs. No tubes.	380
Protected Cruisers.	Buenos Aires	1895	4,500	23	Guns: 2—8 inch, 45 cal.; 4—6 inch, 45 cal.; 6—4.7 inch, 45 cal.; 16—3 pdrs. Torpedo tubes: 3.	400
	Nueve de Julio	1892	3,500	21	Guns: 4—6 inch, 40 cal.; 8—4.7 inch, 40 cal.; 12—3 pdrs. Torpedo tubes: 5.	320
	25 de Maio	1890	3,200	21	Guns: 2—8.2 inch; 8—4.7 inch, 40 cal.; 12—3 pdrs.; 12—1 pdrs. Torpedo tubes (18 inch): 3.	300
	Patagonia .	Reconstructed 1909	1,442	14	Guns: 1—6 inch; 1—4.7 inch; 8—3 pdrs. No torpedo tubes.	210

ARGENTINA.

TORPEDO BOAT DESTROYERS AND TORPEDO CRAFT.

DESTROYERS.

		Displacement (tons)	Speed (knots)	Armament	Complement
4 *Santiago class*	San Luis, Santa Fé, Santiago, Tucuman .	—	32	Guns : 3—4 inch.	—
2 *Cordoba class.*	Cordoba, La Plata .	820	34	Guns : 4—4 inch.	110
2 *Catamarca class.*	Catamarca, Jujuy .	440	34	Guns : 4—4 inch.	110
4 *Mendoza class.*	Mendoza, Rioja, Salta, San Juan . . .	450	32	Guns : 4—4 inch.	110
3 *Corrientes class.*	Corrientes, Entre-Rios, Misiones . . .	280	27	Guns : 1—14 pdr.	54
TORPEDO BOATS					
2 *Commodore Py class.*	Commodore Py . . Muratore . . .	110	24–5	3—3 pdrs	27
6 *Bathurst class.*	Bathurst, Buchardo, Jorge, King, Pinedo, Thorne . . .	85	24	2—3 pdrs	15

BRAZIL.

		Completed	Displacement (tons)	Speed (knots)	Armament	Complement
Dreadnoughts.	Minas Geraes . . Sao Paulo . .	1910	19,281	21	Guns: 12—12 inch, 50 cal; 22—4.7 inch; 8—3 pdrs. [illegible] torpedo tubes.	900
Coast Battleships.	Deodoro . . Floriano . .	1902	3,162	14	Guns: 2—9.2 inch, 45 cal.; 4—4.7 in., 50 cal.; 6—6 pdrs. 4—1 pdrs. Torpedo tubes (18 inch): 2.	200
Cruisers. .	Bahia . . Rio Grande do Sul	1910	3,100	26.5	Guns: 10—4.7 inch; 6—3 pdrs. Torpedo tubes (18 inch): 2.	
Protected Cruisers.	Barrozo . .	1897	3,450	20	Guns: 6—6 inch, 50 cal.; 4—4.7 inch, 50 cal.; 10—6 pdrs.; 4—1 pdrs. 4 Maxims.	300
	Tammandare .		4,537	17	Guns: 10—6 inch; 2—4.7 inch.	450

BRAZIL.

TORPEDO DESTROYERS

		Displacement (tons)	Speed (knots)	Armament	Complement
10 *Para class.*	Amazonas, Matto Grosso, Piauhy, Para Rio Grande der Norte, Parahyba, Alajoas, Santa Catharina, Parana, Sergipe	550	27	Guns : 2—4 inch ; 4—3 pdrs. Torpedo tubes : (18 inch) : 2.	

TORPEDO BOATS.

	Goyaz	—	27	Guns : 2—3 pdrs.	—
3 *Silvado class.*	Silvado, Pedro-Ivo, Pedro-Affonso	130	28	Guns : 2—1 pdrs. 3 tubes	24

SUBMARINES.

3 Laurenti (F.I.A.T.) 1912).

BULGARIA.

TORPEDO GUNBOAT.

Nadiedja. Displacement : 715 tons. Speed : 17 kts. Armament : Guns : 2—4 inch ; 2—9 pdrs. ; 2—3 pdrs. Torpedo tubes : 2.

TORPEDO BOATS.

6 *Creusot* boats (1907) : Smeli, Khrabry, Drsky, Schumni, Lietiachty, Strogi. Displacement : 98 tons. Armament : 2—3 pdrs. Torpedo tubes (18 inch) : 3. Designed H.P. : 2000= 26 kts. 1 Screw. Du Temple boilers.

2 old spar launches. 7 other launches.

CHILI.

		Completed	Displacement (tons)	Speed (knots)	Armament	Complement
1888 Old Battleship.	Capitan Prat	Reconstructed 1909	6,901	19.5	Guns: 4—9.4 inch, 35 cal.; 8—4.7 inch; 6—6 pdrs.; 4—3 pdrs.; 10—1 pdrs. 5 Maxims.	480
1896 Armoured Cruiser.	O'Higgins .	1898	8,500	19	Guns: 4—8 inch, 45 cal.; 10—6 inch; 4—4.7 inch, 45 cal.; 10—12 pdrs.; 2 Pompoms. Torpedo tubes (18 inch): 2.	500
1895 Belted Cruiser.	Esmeralda	1897	7,000	22.25	Guns: 2—8 inch, 40 cal.; 16—6 inch, 40 cal.; 8—12 pdrs.; 2—3 pdrs. 8 Maxims. Torpedo tubes (18 inch): 3.	500
Protected Cruisers.	Chacabuco	1902	4,300	24	Guns: 2—8 inch, 45 cal.; 10—4.7 inch; 12—12 pdrs.; 6—2½ pdrs.; 4 Pompoms. Torpedo tubes: 5.	400
	Ministro Zenteno	1898	3,600	20	Guns: 2—6 inch, 45 cal.; 6—6 inch, 40 cal.; 10—6 pdrs.; 4—1 pdrs.; 4 Maxims. Torpedo tubes: 2.	280
	Blanco Encalada	1894	4,420	22.75	Guns: 2—8 inch, 40 cal.; 10—6 inch, 40 cal.; 12—3 pdrs. Torpedo tubes: 5.	427
	Presidente Errazuriz	Reconstructed 1908	2,100	19	Guns: 4—6 inch, 45 cal.; 2—4.7 inch, 45 cal.; 4—6 pdrs.; 4—1 pdrs. 2 Machine. Torpedo tubes (14 inch): 3.	171

CHILI.

2 TORPEDO GUNBOATS.

Tomé, Talcahuano. Displacement : 790 tons. Speed : 20 kts Armament : Guns : 6—6 pdrs., 50 cal. Torpedo tubes (14 inch) : 5 above water.

11 DESTROYERS.

4 *Lynch class :* Almirante Lÿnch, Al. Condell, Al. Recolledo, Al. Riceron (1911-12). Displacement : 1,430 tons. H.P. : 27,000. Speed : 32 kts. Complement : —. Torpedo tubes : —.

1 Capitan Thomson. Displacement : 480 tons. Speed : 31 kts. Torpedo tubes : 2.

2 *Capitan O'Brien class :* Capitan O'Brien, Capitan Merino Jarpa (1902). Displacement : 300 tons. H.P. : 6,250. Speed : 30 kts. Complement : 65. Torpedo tubes : 2.

4 *Capitan Orella class :* Capitan Orella, Capitan Munoz Gamero, Teniente Serrano, Guardia Marina Riquelme (1896). Displacement : 300 tons. H.P. : 6,000. Speed : 30 kts. Complement : 65 Torpedo tubes : 2.

5 TORPEDO BOATS.

5 *Ingeniero Hyatt class :* Ingeniero Hyatt, Cirujano Videla, Ingeniero Mutilla, Guardia Marina Contreras, Teniente Rodriguez (1896-98). Displacement : 140 tons. H.P. : 2,200. Fuel : 40 tons. Complement : 28. Armament : 3—3 pdrs. Torpedo tubes (14 inch) : 3.

CHINA.

		Completed	Displacement (tons)	Speed (knots)	Armament	Complement
Protected Cruisers.	One . / Three .	Building	4,900 / 1,900	32	Guns: 10—4 inch Skoda. Torpedo tubes: 2.	
	Ying Swei / Chao-Ho	1912	2,750	22	Guns: 2—6 inch; 4—4 inch; 4—12 pdrs.; 6—3 pdrs. Torpedo tubes (18 inch): 2.	270
	Hai-Chi .	1899	4,300	24	Guns: 2—8 inch; 10—4.7 inch; 12—3 pdrs.; 3—1 pdrs.; 6 Maxims. Torpedo tubes (18 inch): 5.	450
	Hai-Yung / Hai-Chu / Hai-Chen	1898	3,000	19.5	Guns: 3—6 inch, 40 cal.; 8—4.1 in.; 6—1 pdrs.; 6 Max. Torpedo tubes (18 inch): 1 sub.	244
	Fu Hung	[illegible]	2,600	22		
Torpedo Gunboats.	Kien-Wei / Kien-An	1900 / 1901	871	23	Guns: 1—4 inch; 3—9 pdrs.; 6—1 pdrs. Torpedo tubes (14 inch): 2.	300
	Fei-Ying	1897	850	22	Guns: 2—4.1 inch; 2-6 pdrs.; 2-1 pdrs.	90
River Gunboat. .	———		140	12	Guns: 1—24 pdr. Howitzer.; 2—3 pdrs.; 2 Machine.	

CHINA.

TORPEDO CRAFT.

DESTROYERS.

1 *Stabilimento Technico :* Lung-Tuan.
3 *Schichau :* Fupo, Yen-yun, Chang-yen (1912). Armament : 2—12 pdrs., 4—3 pdrs. Speed : **32 kts.**

TORPEDO BOATS.

4 *Shen class :* Shen, Su, Nia, Tchung. Displacement : 97 tons. Armament : 2—3 pdrs. Torpedo tubes : 3. Speed : 23 kts.
2 ———. Speed : 21½ kts.
2 ———. Speed : 23 kts.

COLOMBIA.

		Displacement (tons)	Speed (knots)	Armament
Light Cruiser	Almirante Lezo	1,200	16	Guns: 2—4.7 inch; 4—4 inch Torpedo tubes (14 inch): 4.
Gunboats	General Pinzon	740	—	
	Bolivar	981	—	Guns: 2—3 pdrs.
	Bogota	—	—	Guns: 2—3 pdrs.

CUBA.

		Displacement (tons)	Speed (knots)	Armament
Light Cruiser	Cuba	2,055	18	Guns: 2—4 inch; 4—3 pdrs.
Gunboats	Patria	1,100	—	Guns: 2—12 pdrs.
	10 De Otoubre / 24 De Febrero	165	10.5	Guns: 1—1 pdr.
	Baire	500	14	Guns: 2—6 pdrs.; 2—3 pdrs.
	Hatuey	540	13	Guns: 2—6 pdrs.; 2—3 pdrs.; 2—1 pdrs.
	Yara	450	12	Guns: 2—6 pdrs.; 2—1 pdrs.
	E. Villuendas	180	16	Guns: 2—3 pdrs.

DENMARK.

		Completed	Displacement (tons)	Speed (knots)	Armament	Complement
Coast Service Battleships.	Niels Juel . Peder Skram	1908	3,735	16	Guns : 2—9.4 inch, 45 cal. ; 4—6 inch, 50 cal. ; 10—12 pdrs. ; 2—1 pdrs. 3 Machine. Torpedo tubes : 4.	257
	Olfert Fischer Herluf Trolle	1905 1901	3,650	15.8	Guns : 2—9.4 inch, 40 cal. ; 4—6 inch ; 12—6 pdrs. ; 2 Machine. Torpedo tubes : 3.	255
	Skjold .	1899	2,195	13.4	Guns : 1—9.4 inch ; 3—4.7 inch ; 4—6 pdrs. ; 2 Machine.	137
Old Battleship.	Iver Hvitfeldt	1889	3,446	15.5	Guns : 2—10 inch ; 6—3 pdrs. ; 10 Machine. Torpedo tubes : 2.	265
Cruisers.	Hejmdal .	1896	1,313	17	Guns : 2—4.7 inch ; 4—3.5 inch ; 4—6 pdrs. ; 3 Machine. Torpedo tubes : 4.	156
	Gejser .	1893	1,282	17.1	Guns : 2—4.7 inch ; 4—3.5 inch ; 7 Machine. Torpedo tubes : 4.	156
	Valkyrien .	1888	3,020	17.4	Guns : 2—8.2 inch, 18 cal. ; 6—6 inch, 32 cal. ; 6 Machine. Torpedo tubes : 5.	282

DENMARK.

TORPEDO BOATS.

3 *Yarrow type:* Söridderen, Söulven, Flyvefisken. Speed: 27·5 kts. Armament: 2—12 pdrs. Complement: 33.

3 *Schichau type:* Tumieren, Vindhunden, Spækhuggeren. Speed: 27·5 kts. Armament: 2—12 pdrs. Complement: 33.

3 *Havörnen class:* Havörnen, Hajen, Söbjörnen. Speed: 23·5 kts. Armament: 1—3 pdr. Complement: 24.

2 *Makrelen class:* Makrelen, Nordkaperen. Speed: 20 kts. Armament: 2 Machine guns. Complement: 20.

2 *Havhesten class:* Havhesten, Narhvalen. Speed: 21·2 kts. Armament: 2 Machine guns. Complement: 20.

2 *Stören class:* Stören, Sölöven. Speed: 22 kts. Armament: 2 Machine guns. Complement: 20.

1 Ormen. Speed: 26 kts. Complement: 21.

1 Springeren. Speed: 18 kts. Complement: 17.

SUBMARINES.

8 *Whitehead type:* Havmanden (Fiume) and Havfruen, Thetis, Najaden, Nymfen, 2 April, Neptune, Galatea. Speed: 13-10 kts. Torpedo tubes: 2.

1 *Fiat San Giorgio type:* Dykkeren. Speed: 12-7½ kts. Torpedo tubes: 2 (2 torpedoes to each). Complement: 9.

ECUADOR.

		Completed	Displacement (tons)	Speed (knots)	Armament	Complement
Cruisers .	Libertador Bolivar	1896	800	21	Guns : 2—4.7 inch ; 4—3 pdrs.; 2 Max. Torpedo tubes (18 inch) : 3.	—
	Ex Papin . .	1886	811	12	Guns : 2—5.5 inch ; 1—4 inch ; 5 small.	—

EGYPT.

Khedive's Yacht. .	Mahroussa	1906	3,417	17.5	—	
Customs Cruisers.	Abdul Moneim	1902	598	13	Guns : 1—3 pdr.	
	Abbas .	1891	298	13	Guns : 1—3 pdr.	
	Nour-el-bahr	1884	450	13	Guns : 1—3 pdr.	
	Reschid .		650	—	Guns : 2—6 pdrs.	
Ports and Lights Administration.	Aida . .	—	723	—	1 gun.	

GREECE.

		Completed	Displacement (tons)	Speed (knots)	Armament	Complement
Dreadnought	Unnamed .	Ordered	23,500	22	Guns: 10—13.4 inch; 22—5.5 inch. Torpedo tubes: 4.	
Battle Cruiser.	George I (ex *Salamis*)*	Due 1915	19,500	23	Guns: 8—14 inch, 45 cal.; 12—6 inch, 50 cal. Torpedo tubes: —.	
Armoured Cruiser.	G. Averoff	1911	10,118	22.5	Guns: 4—9.2 inch, 45 cal.; 8—7.5 inch, 45 cal.; 16—12 pdrs.; 2 Field, 12 pdrs. Torpedo tubes (18 inch): 3.	550
		(Recon-	structed)			
Old Battleships. *Spetsai class.*	Psara .	1897	5,000	17	Guns: 2—10.8 inch, 36 cal.; 1—10.8 inch, 30 cal.; 5—6 inch, 45 cal.; 1—4 inch, 50 cal.; 8—9 pdrs.; 12—1 pdrs. No torpedo tubes.	440
	Spetsai .	1899				
	Hydra ..	1900				
Light Cruisers.	Ad. Coundourietis	Due 1916	5,500	26	Guns: 8—6 inch. Torpedo tubes: 2	
	"X" . .	1917				
Light Cruiser.	Helle .	1913	2,600	20	Guns: 2—6 inch; 4—4 inch; 2—14 pdrs.; 6—3 pdrs.; 2—1 pdrs. Torpedo tubes (18 inch): 2.	232
Torpedo Depot Ship.	Kanaris .	1877	1,100	14	Guns: 2—4 inch. Torpedo tubes 4.	

*Building in Germany—May be taken over by the German Navy.

GREECE.

DESTROYERS.

4 *Laird :* Aetos, Leon, Jerax, Panthir. Speed : 32 kts. Armament : 4—4 inch. Complement : 32.

2 *Vulkan :* Keravnos, Neagenea. Speed : 32.5 kts. Armament : 4—4 inch.

4 *Thyella class :* Thyella, Sfendoni, Naphkratoussa, Lonchi. Speed : 30 kts. Armament : 2—12 pdrs. Complement : 58.

4 *Doxa class :* Aspis, Niki, Doxa, Velos. Speed : 30 kts. Armament : 2—12 pdrs. Complement : 58.

TORPEDO BOATS.

2 *Vulkan :* Aigli, Thetis, Daphne, Arethusa, Alcyon, Doris. Speed : 25 kts. Armament : 2—6 pdrs.

Nikopolis. Speed : 27 kts. Armament : 2—6 pdrs.

SUBMARINES.

Delphin, Xiphias. Speed : 14·9 kts. Torpedo tubes (18 inch) : 5.

HOLLAND.

		Completed	Displacement (tons)	Speed (knots)	Armament	Complement
Battleships.	One. . .	Building	7,600	18	Guns: 4—11 inch; 10—4 inch. Torpedo tubes (21 inch) 3.	—
	Zeven Provincien	1910	6,530	16	Guns: 2—11 inch, 40 cal.; 4—6 inch, 40 cal.; 10—12 pdrs. 55 cal.; 4—1 pdrs. Torpedo tubes (18 inch): 2.	411
	Heemskerck .	1908	5,130	16	Guns: 2—9.4 inch, 45 cal.; 6—5 inch, 40 cal.; 6—12 pdrs.; 4—1 pdrs. Torpedo tubes: 2.	355
	Tromp . .	1906	5,300	16.5	Guns: 2—9.4 inch, 40 cal.; 4—6 inch, 40 cal.; 10—12 pdrs.; 4—1 pdrs. Torpedo tubes (18 inch): 3.	344
	Koningin Regentes De Ruyter Hertog Hendrik	1902 1904 1903	4,950	16	Guns: 2—9.4 inch, 40 cal.; 4—6 inch, 40 cal.; 8—12 pdrs.; 2—1 pdrs. Torpedo tubes (18 inch): 3	320
	Evertsen . . Piet Hein. . Kortenaar .	1896	3,520	16	Guns: 3—8.2 inch, 35 cal.; 2—6 inch; 6—12 pdrs.; 6—1 pdrs. Torpedo tubes: 2.	265

HOLLAND.

		Completed	Displacement (tons)	Speed (knots)	Armament	Complement
Cruisers. *Holland class.*	Gelderland . .					
	Holland . .	1895	3,950	20	Guns: 2—6 inch, 40 cal.; 6—4.7 inch; 4—12 pdrs.; 8—1 pdrs.; 4 Machine. Torpedo tubes: 2.	333
	Zeeland . .					
	Noordbrabant .	1897				
Coast Defence Monitors.	Brinio . .					
	Frisco . .	1912	540	16	Guns: 4—4.1 inch; 2 Machine.	
	Gruno . .					
	Reinier-Claeszen	1891	2,490	12.5	Guns: 1—8.2 inch, 35 cal.; 1—6.7 inch, 35 cal.; 4—6 pdrs.; 3—1 pdrs. Torpedo tubes (14 inch): 2.	160
Gunboats.	Assahan . .					
	Koetei . .	1900	830	13	Guns: 2—4.1 inch; 2—12 pdrs.	105
	Siboga . .	1898				
	Edi . .					
	Mataram . .	1896	810	13	Guns: 3—4.7 inch.	95
	Serdang . .	1896	600	12.5	Guns: 3—4.7 inch; 1—12 pdr.	
Mine Ships.	Dra, Medusa .	1911	670	11.5	Guns: 3—12 pdrs.	
Fishery Cruisers.	Zeehond . .	1892	710	—	Guns: 6—4.1 inch.	
	Dolfijn . .	1889	250	12	Guns: 2—12 pdrs.	
Gunnery Tender.	Bellona . .	1892	930	8		

HOLLAND.

DESTROYERS.

Fret class: Fret, Wolf, Jakhals, Bulhond, Hermyln, Lynx, Panter, Vos. Speed: 30 kts. Armament: 4—12 pdrs.; 2 Machine. Complement: 84.

TORPEDO BOATS.

4 *New 180 ton boats* (1912-13). G 16—G 13. Armament: 2—12 pdrs. Torpedo tubes (18 inch): 3.
4 ('07-'09 boats): G 9, 10 and 12, and *with two masts*, G 11. Armament: 2—3 pdrs.
4 ('05-'08 boats): G 8—5. Armament: 2—3 pdrs.
4 ('04-'06 boats): G 4—1. Armament: 2—3 pdrs.
3 ('02-'04 boats): Tangka, Smeroe, Wajang. Armament: 2—3 pdrs.
3 *Ophir class:* Ophir, Pangrango, Rindjani. Armament: 2—3 pdrs.
3 *Draak class:* Draak, Krokodil, Zeeslang. Armament: 2—3 pdrs.
3 *Sphinx class:* Minotaurus, Python, Sphinx. Armament: 2—3 pdrs.
2 *Hydra class:* Hydra, Scylla. Armament: 2—3 pdrs.
5 *Foka class:* Empong, Foka, Goentoer, Habang, Idjen. Armament: 2—1 pdrs.
3 *Ardjoeno class:* Ardjoeno, Cerberus, Cycloop. Armament: 2—1 pdr.
3 *Ex M. Gardijn class:* K 1—K3. Armament: 2—1 pdr.
3 *Nobo class:* Lamogan, Makjan, Nobo. Armament: 2—1 pdr.
3 *J Hobein class:* Jacob Hobein, Jan Haringh, Jasper Lijnsen. Armament: 2—1 pdrs.

SUBMARINES.

4—6 & 7, K 2 & 3 (w). Speed: $\frac{16}{11}$ kts.
3—4 & 5 (w) & K 1. Speed: $\frac{16}{11}$ kts. Torpedo tubes: 2.
2—2 & 3 (w). Speed: $\frac{16}{11}$ kts. Torpedo tubes: 2.
1—No. 1. Speed: $\frac{9}{7}$ kts. Torpedo tubes: 1. Complement: 9.

ITALY.

	Completed	Displacement (tons)	Speed (knots)	Armament	Complement
1914 Super-Dreadnoughts. Cristoforo Colombo, Marcantonio Colonna, Francesco Morosini, Caracciolo	Building	30,000	25	Guns: 8—15 inch; 16—6 inch; 20—3 inch.	
1912 Dreadnoughts. Caio Duilio, Andrea Doria	1915	22,000	22.5	Guns: 13—12 inch, 46 cal.; 16—6 inch; 18—3 inch. Torpedo tubes (18 inch): 3.	957
1910 Dreadnoughts. Conte di Cavour, Giulio Cesare, Leonardo Da Vinci	1914	22,000	22.5	Guns: 13—12 inch, 46 cal.; 18—4.7 inch; 14—3 inch.	957
1909 Dreadnought. Dante Alighieri.	1912	18,400	22.5	Guns: 12—12 inch, 46 cal.; 20—4.7 inch; 12—3 inch. Torpedo tubes: 3.	900
San Giorgio class. San Giorgio, San Marco	1910	9,830	23	Guns: 4—10 inch, 45 cal.; 8—7.5 inch, 45 cal.; 18—12 pdrs.; 2 Field, 12 pdrs. Torpedo tubes (18 inch): 3.	550
Pisa class. Pisa, Amalfi	1909	10,118	22.5	Guns: 4—10 inch, 45 cal.; 8—7.5 inch, 45 cal.; 16—14 pdrs.; 2 Field, 12 pdrs. Torpedo tubes: (18 inch) 3.	550

ITALY.

		Completed	Displacement (tons)	Speed (knots)	Armament	Complement
Vittorio Emanuele class.	Vittorio Emanuele III. Regina Elena . Napoli . . Roma . .	1907 1908	12,625	21	Guns: 2—12 inch, 40 cal.; 12—8 inch, 45 cal.; 24—12 pdrs.; 2—3 pdrs. 2 Maxims. Torpedo tubes (18 inch): 2.	1000
Brin class.	Benedetto Brin . Regina Margherita	1904	13,427	18	Guns: 4—12 inch, 40 cal.; 4—8 inch; 12—6 inch; 20—12 pdrs.; 2—1 pdrs.; 2 Maxims. Torpedo tubes (18 inch): 4.	720
G. Garibaldi class.	Giuseppe Garibaldi Varese . . Francesco Ferruccio	1901 1904	7,400	20	Guns: 1—10 inch, 45 cal.; 2—8 inch, 45 cal.; 14—6 inch, 40 cal.; 10—12 pdrs.; 6—3 pdrs.; 2—1 pdrs. Torpedo tubes (18 inch): 4.	517
St. Bon class.	Ammiraglio Di St. Bon . . Emanuele Filiberto	1901	9,800	16	Guns: 4—10 inch, 40 cal.; 8—6 inch, 40 cal.; 8—4.7 inch (*St. Bon*); 8—6 pdrs. (*Filiberto*); 6—12 pdrs.; 6—3 pdrs. Torpedo tubes (18 inch): 4.	542
	Carlo Alberto . Vettor Pisani .	Refitted 1904	6,500	17	Guns: 12—6 inch, 40 cal.; 6—4.7 inch, 40 cal.; 14—6 pdrs.; 2—1 pdrs. Torpedo tubes: 4.	500
	Marco Polo .	1895	4,583	19	Guns: 6—6 inch, 40 cal.; 4—4.7 inch; 6—6 pdrs.; 2—1 pdrs.; 2 Machine. Torpedo tubes: 4.	394

ITALY.

		Completed	Displacement (tons)	Speed (knots)	Armament	Complement
Old Armoured Ships.	Sicilia .	—	13,375	20	Guns: 4—old 13.5 inch; 8—old 6 inch; 16—4.7 inch; 20—6 pdrs. Torpedo tubes: 6.	783
	Sardegna		13,860			
1911 Scout Cruisers.	Marsala .	1913	3,470	28	Guns: 6—4.7 inch; 6—3 inch. Torpedo tubes (18 inch): 2.	197
	Nino Bixio					
	"D" .	1914				
1911 Scout Cruiser.	Quarto .	1912	3,300	28	Guns: 6—4.7 inch; 6—3 inch. Torpedo tubes (18 inch): 2.	197
Protected Cruisers.	Libia (ex *Drama*)	1913	3,800	22	Guns: 2—6 inch, 45 cal.; 8—4.7 inch, 50 cal. Torpedo tubes (18 inch): 2.	276
	Puglia .	1901	2,538	18.5	Guns: 6—4.7 inch, 40 cal.; 6—6 pdrs.; 2—1 pdrs.; 2 Machine; 1—12 pdr. Field. Torpedo tubes (18 inch): 2.	274
	Calabria	1894	2,492	16	Guns: 4—6 inch; 6—4.7 inch; 6—6 pdrs.; 2—1 pdrs.; 2 Machine. Torpedo tubes (14 inch): 2.	254

ITALY.

		Completed	Displacement (tons)	Speed (knots)	Armament	Complement
Protected Cruisers.	Elba (Balloon Ship)	1895	2,730	18—5	Guns: 6—4.7 inch, 40 cal.; 4—6 pdrs.; 2—1 pdrs.; 1—12 pdr.. Field. Torpedo tubes (14 inch): 2.	247
	Lombardia. Eturia. Liguria.	1891 to 1894	2,280	18.5	Guns: 6—4.7 inch, 40 cal.; 6—6 pdrs.; 2—1 pdrs. Torpedo tubes (14 inch): 3.	246
	Piemonte	1889	2,650	23	Guns: 10—4.7 inch; 6—6 pdrs.; 2—1 pdrs.; 2 Maxims. Torpedo tubes (14 inch): 2.	325
	Etna	1887	3,350	17.8	Guns: 4—6 inch; 2—4.7 inch; 2—12 pdrs.; 4—6 pdrs.; 2—3 pdrs. Torpedo tubes (14 inch): 3.	308
	Giovanni Bausan	1884	3,330	17.5	Guns: 2—old 10 inch; 4—6 inch; 4—6 pdrs.; 2—1 pdrs.; 2 Maxims. Torpedo tubes (14 inch): 3.	
	Agordat. Coatit.	1901	1,313	23	Guns: 12—12 pdrs. Torpedo tubes (18 inch): 2.	154
Torpedo Gunboats. (Mostly fitted as Mine-Layers.)	Minerva. Partenope, Iride	—	846	22	Guns: 2—12 pdrs.; 4—6 pdrs.; 3—1 pdrs. Torpedo tubes: 6.	105
	Montebello	—	860	20	Guns: 4—6 pdrs. Torpedo tubes: 3	105
	Goito. Tripoli.	—	840	20	Guns: 6—6 pdrs. Torpedo tubes: 3	

ITALY.

DESTROYERS.

3 *Ansaldo :* C. Mirabello, C. A. Racchia, A. Riboty. Speed : 32 kts.

3 *Ansaldo :* A. Poerio, C. Rossardi, G. Pepe.

6 *Odero :* R. Pilo, G. Abba, I. Nievo, S. Schiaffino, P. Bronzetti, G. Missori.

4 *Pattison (Thornycroft design) :* F. Nullo, A. Mosto, G. Sirtori, G. Carini.

2 *Orlando :* Ardito, Ardente. Speed : 30 kts. Armament : 1—4.7 inch.

2 *Orlando :* Animoso, Audace. Armament : 1—4.7 inch.

6 *Pattison :* Indomito, Impavido, Impetuoso, Insidioso, Intrepido, Irrequieto. Armament : 1—4.7 inch.

10 *Ansaldo :* Artigliere, Alpino, Corazziere, Fuciliere, Granatiere, Lanciere, Bersagliere, Pontieri, Carabiniere, Garibaldino. Armament : 4—12 pdrs.

8 *Nembo (Thornycroft pattern) :* Aquilone, Borea, Espero, Nembo, Turbine, Zeffiro. Armament : 5—6 pdrs.

6 *Schichau :* Dardo, Euro, Lampo, Strale, Ostro. Armament : 1—12 pdr.

1 *Ansaldo :* Ascaro.

1 *Ansaldo :* Fulmine. Armament : 1—12 pdr.

ITALY.

TORPEDO BOATS.

12 128—68. Speed : 20 kts. Complement : 17.

6 56—29 (third class). Speed : 21 kts. Complement : 11.

1 Gabbiano. Speed : 22½ kts. Complement :

1 Pellicano. Speed : 23 kts. Complement : 28.

1 Condore. Speed : 25.7 kts. Complement : 28.

6 *Saffo class :* Saffo, Serpente, Sagittario, Scorpione, Sirio, Spica. Speed : 25.5 kts. Armament : 3—3 pdrs. Complement : 34.

6 *Airone class :* Airone, Alcione, Ardea, Albatros, Astore, Arpia. Speed : 26 kts. Armament : 3—3 pdrs. Complement : 34.

8 *Cigno class :* Cigno, Clio, Canopo, Centauro, Cassiopea, Calliope, Calipso, Climene. Speed : 26 kts. Armament : 3—3 pdrs. Complement : 34.

4 *Pegaso class :* Pegaso, Perseo, Procione, Pallade. Speed : 26 kts.

4 *Orsa class :* Orsa, Orione, Orfeo, Olimpia. Speed : 26 kts. Armament : 3—3 pdrs. Complement : 34.

2 *Aquila class :* Avvoltoio, Sparviero. Speed : 23.1 kts. Armament : 2—1 pdrs. Complement : 23.

2 32—31.

6 30—25.

12 24—13.

12 12—1.

ITALIAN "DANTE ALIGHIERI" (Battleship).

ITALIAN "SAN MARCO" (Cruiser). *Silk*

ITALY.

SUBMARINE FLOTILLA.

6 new large submarines : P. Micca, Galvani, Toricelli, L. Mocenigo, L. Marcello, A. Emo.

8 *Argo class :* Argo, Fisalia, Jalia, Janita, Medusa, Salpa, Velella, Zoca.

2 *G. Pullino class :* G. Ferraris, G. Pullino.

2 *Nautilus class :* Nautilus, Nereide.

5 *Glauco class :* Glauco, Squalo, Narvalo, Tricheco, Otaria.

1 Atropo.

1 Delfino.

MEXICO.

		Completed	Displacement (tons)	Speed (knots)	Armament	Complement
Cruiser Transports.	General Guerrero	1909	1,880	12	Guns: 6—4 inch; 2—3 pdrs.	
	Progreso . .	1907	1,590	14	Guns: 4—4 inch.	
4 Gunboats.	Bravo, Morales .	1905	1,200	16	Guns: 2—4 inch; 6—6 pdrs. Torpedo tubes (14 inch): 1.	116
	Tampico, Vera Cruz	1903	980	16	Guns: 2—4 inch; 6—6 pdrs. Torpedo tubes (14 inch): 1.	98
Old Cruiser. .	Zaragoza . .	Refitted 1910	1,200	15.5	Guns: 6—4.7 inch, 40 cal.; 2—6 pdrs. 2—1 pdrs.	270

PANAMA.

		Completed	Displacement (tons)	Speed (knots)	Armament	Complement
	Padilla . .	—	75	10	Guns: 1—1 pdr.	
	Darian, Gatan .	—	400	15	Guns: 3—1 pdrs.	
	Chercuito .	—	643	12	Guns: 2—1 pdrs.	

PERSIA.

		Completed	Displacement (tons)	Speed (knots)	Armament	Complement
Gunboat. .	Persepolis .	1885	600	10.5	—	
	Muzalfari . .	purchased 1903	379	—	Guns: 2—3 pdrs., Hotchkiss. Several small.	
Imperial Yacht.	Setika. . .	1902	400	—	—	

NORWAY.

		Completed	Displacement (tons)	Speed (knots)	Armament	Complement
Coast Service Battleships.	Bjoergvin . Nidaros .	due 1915	4,900	15	Guns: 2—9.4 inch; 4—6 inch; 6—4 inch. Torpedo tubes (18 inch): 2.	250
	Norge . Eidsvold .	1900	4,200	17	Guns: 2—8.2 inch, 45 cal.; 6—6 inch, 45 cal.; 8—12 pdrs.; 4—3 pdrs. Torpedo tubes: 2.	270
	Harald Haarfarge Tordenskjold	1897	3,900	17.2	Guns: 2—8.2 inch; 6—4.7 inch; 6—12 pdrs.; 2—1 pdrs. Hotchkiss. Torpedo tubes: 2 submerged.	249
2 Monitors.	Thor . Thrudvang	1897	2,000	8.5	Guns: 2—4.7 inch, 40 cal.; 2—9 pdrs.; 3 Machine.	
1st Class Gunboats.	Frithjof .	1898	1,480	15	Guns: 2—4.7 inch; 6—12 pdrs.; 4—3 pdrs. Torpedo tubes (18 inch): 1.	166
	Viking .	1892	1,200	15.5	Guns: 1—4.7 inch; 1—6 inch; 4—12 pdrs.; 6—3 pdrs. Torpedo tubes: 1.	153

NORWAY.

DESTROYERS.

Draug and Troll. Displacement: 580 tons. Armament: 6—12 pdrs. Speed: **27 kts.** Complement: 71. Garm and 2 others building this type.

1 *Schichau type:* Valkyrien (1896). Displacement: 415 tons. Complement: 59. Guns: 4—12 pdrs. Speed: 23 kts.

TORPEDO BOATS.

10 *Delfin class:* Brand, Delfin, Hai, Hval, Laks, Sael, Storm, Sild, Skrei, Trods.

3 *Skarv class:* Skarv, Kjell, Teist.

5 *Grib class:* Grib, Fo, Lovm, Orn, Ravn.

2 *Hauk class:* Falk, Hauk.

2 *Varg class:* Rakct, Varg.

8 *Hvas class:* Blink, Djerv, Dristig, Glimt, Hvas, Kjaek, Kvik, Lyn.

1 Myg.

SUBMARINE.

A 1. Speed: $\frac{1}{8}^{2}$ kts. Complement: 12.

3 more *Germania* boats building. (A 2—A 5.)

PERU.

		Completed	Displacement (tons)	Speed (knots)	Armament	Complement
Armoured Cruiser.	Commandante Aguirre (ex *Dupuy de Lôme*)	1890 Purchased 1911	6,400	22	Guns: 2—7.6 inch; 6—6.4 inch; 4—9 pdrs.; 12—3 pdrs. Torpedo tubes (18 inch): 4.	521
Light Cruisers.	Almirante Grau . Coronel Bologuesi .	1907	3,200	24	Guns: 2—6 inch, 50 cal; 8—14 pdrs.; 8—1 pdrs. Torpedo tubes (18 inch): 2.	300
	Lima	1882	1,700	16.2	Guns: 2—6 inch, Krupp; 2—4.7 in.; 2—1 pdrs.; 2 Max.	—
Destroyer.	Teniente Rodriguez .	1909	500	35		
Submarines. .	Ferré . .	1912	300	14		
	Palacios .	1913	440	9		
	Constitucion -	—	—	—	Guns: 6—3 pdrs.: 2 Maxims.	
	Iquitos . .	—	—	—	Guns: Small Q.F.	
	Chalaio	—	—	—	Guns: 2—3 inch, Krupp (B.L.)	

PORTUGAL.

		Completed	Displacement (tons)	Speed (knots)	Armament	Complement
Battleship.	Vasco Da Gama	Reconstructed 1902	3,030	15.1	Guns : 2—8 inch ; 1—6 inch ; 1—12 pdr. ; 8—3 pdrs. ; 4 Machine. No torpedoes.	259
Cruisers.	Republica (ex *R. D. Amelia*)	1901	1,683	18	Guns : 4—6 inch, 45 cal. ; 2—4 inch, 45 cal. ; 4—3 pdrs. ; 2—1 pdrs. Torpedo tubes (18 inch) : 2.	263
	Almirante Reis (ex *Dom Carlos I*)	1899	4,253	22	Guns : 4—6 inch, 45 cal. ; 8—4.7 inch, 45 cal. ; 14—3 pdrs. ; 2—1 pdrs. ; 3 Machine. Torpedo tubes (18 inch) : 5.	418
	Sao Gabriel	1899	1,800	17	Guns : 2—6 inch ; 4—4.7 inch ; 8—3 pdrs. ; 2—1 pdrs. ; 3 Machine. Torpedo tubes (18 inch) : 1.	242
	Adamastor	1897	1,757	18	Guns : 2—6 inch ; 4—4.7 inch ; 4—6 pdrs. ; 2—1 pdrs. ; 3 Machine. Torpedo tubes (18 inch) : 3.	235
Destroyers	6 "X"			32		
	2 Duro (Y)			27*t*		73
	Tejo			25		85
4 Torpedo Boats.	No. 2, 3, 4 (Y)			19		18
	No. 1 (Y)			19		15

1 Submarine. *Fiat type*. Torpedo tubes : 2. (3 somewhat larger *pro.*)

ROUMANIA.

		Displacement (tons)	Speed (knots)	Armament
Cruiser.	Elizabetha	1,320	18.5	Guns : 4—6 inch ; 4—6 pdrs. ; 2—1 pdrs. Torpedo tubes (18 inch) : 4.
Monitors.	Joan Bratiano, Laskar Catargi, Mihail Cogalnicheanou, Alexander Lahovari .	600	14	Guns : 3—4.7 inch ; 4—3 pdrs. ; 2—4.7 inch Howitzers.

DESTROYERS.

4 (Pattison, Naples). Speed : 35 kts. Armament : 3—4 inch ; 7—12 pdrs.

TORPEDO BOATS.

3 ***Naluca class :*** Naluca, Shorul, Smeul. Speed : 21 kts.

SIAM.

		Displacement (tons)	Speed (knots)	Armament
	Maha Chakrkri	3,000	14	Guns: 4—4.7 inch; 10—6 pdrs.; 6 Machine.
Gunboats	Bhali Sukrib. Murathia	530-580	12	Guns: 1—4.7 inch; 5—6 pdrs.; 4—1 pdrs.
	Mahut Rajakumar	700	11.5	Guns: 2—4.7 inch; 4—6 pdrs.; 3—1 pdrs. 1 tube.
Customs Cruiser	Sriya Monthon	—	14.5	Guns: 1—6 pdrs.
Transport	Buk	850	10	—

DESTROYERS.

Sua-Tayaru-Schol (1908), *B* and *C building*. Displacement: 375 tons. Armament: 1—12 pdr.; 5—6 pdrs.; 2 Maxims. Torpedo tubes (18 inch): 2. H.P.: 6000=27 kts.

TORPEDO BOATS.

Kawasaki boats: No. 1, 2 and 3 (1908). Displacement: 120 tons. Armament: 1—6 pdr.; 1—2½ pdr. Torpedo tubes (18 inch): 2. Speed: 22 kts.

SPAIN.

		Completed	Displacement (tons)	Speed (knots)	Armament	Complement
1909 Dreadnoughts.	Jaime I .	1915	15,700	19.5	Guns: 8—12 inch, 50 cal.; 20—4 inch; 2—3 pdrs.; 2 Maxims.	700
	Alfonzo XIII	1914				
	Espana .	1913				
Armoured Ships.	Pelayo .	Reconstructed 1897	9,950	16	Guns: 2—12.6 inch, 35 cal.; 2—11 inch, 35 cal.; 9—5.5 inch, 45 cal.; 12—6 pdrs.; 9—1 pdrs.; 2 Mach. Torpedo tubes: 7.	621
Cataluna class.	Princesa de Asturias .	1902	7,000	18	Guns: 2—9.4 inch; 8—5.5 inch; 2—12 pdrs. (boat); 8—6 pdrs.; 8—3 pdrs. Torpedo tubes: 2.	497
	Cataluna .	1903				
Protected Cruisers.	Reina Regente	1908	5,372	21	Guns: 10—6 inch; 12—6 pdrs.; 8—3 pdrs.; 2—3 inch Field.	497
	Estramadura	1902	2,030	18	Guns: 8—4 inch; 2—6 pdrs.; 4—1 pdrs.; 4—12 pdrs. (boat.)	260
	Rio de la Plata	1899	1,713	20	Guns: 2—5.5 inch, 35 cal.; 4—4 inch; 6—6 pdrs.; 4—1 pdrs.	213
	Emperador Carlos V .		9,200	20	Guns: 2—9.4 inch; 8—5.5 inch; 4—4 inch; 2—12 pdrs. (boat); 8—6 pdrs.; 8—1 pdrs.; 2 Machine. Torpedo tubes (14 inch): 2.	600

*Owing to the war, a new Spanish programme has been countermanded, and held over pending naval developments.

SPAIN.

DESTROYERS.

Bustamente class: Bustamente, Villamil, Cadarso. Speed: 28 kts. Armament: 5—6 pdrs.

Audaz class: Audaz, Osado; also Proserpina and Terror. Speed: 30 kts. Armament: 2—14 pdrs.; 2—6 pdrs. Complement: 67.

1 Proserpina. Speed: 30 kts. Complement: 70.

1 Terror. Speed: 28 kts. Complement: 67.

TORPEDO BOATS.

2 *Acevedo class:* Acevedo, Julian Ordonez.

24 New. Speed: 29 kts. Armament: 3—6 pdrs.

1 Halcon. Speed: 24 kts. Armament: 4—3 pdrs.

1 Habana. Speed: 21 kts. Armament: ——.

1 Orion. Speed: 21 kts. Armament: 2—1 pdr.

SUBMARINES.

3 of 600 tons (sub.) to be built at Cartagena, 1915-16.

SWEDEN.

		Completed	Displacement (tons)	Speed (knots)	Armament	Complement
	Sverige . Gustav V . Drottning . Wictoria .	1915 1916 1917 1918	7,000	22.5	Guns: 4—11 inch, 45 cal.; 8—6 inch, 50 cal.; 6—12 pdrs. Torpedo tubes: 2.	408
	Oscar II .	1907	4,700	18.3	Guns: 2—8.2 inch, 45 cal.; 8—6 inch, 50 cal.; 10—6 pdrs.; 3—1 pdrs. Torpedo tubes (18 inch): 2.	326
Aran class.	Aran . Vasa . Tapperheten Manligheten	1902 1903 1904 1906	3,700	17	Guns: 2—8.2 inch, 45 cal.; 6—6 inch; 10—3 pdrs.; 2—1 pdrs. Torpedo tubes (18 inch): 2.	250
Battleships.	Dristigheten	1901	3,600	16.8	Guns: 2—8.2 inch, 45 cal.; 6—6 inch, 45 cal.; 10—6 pdrs.; 2—1 pdrs. Torpedo tubes: 2.	250
	Thor . Niord .	1899 1900	3,700	16.5	Guns: 2—10 inch, 42 or 45 cal.; 6—4.7 inch, 45 cal.; 10—6 pdrs.; 2—1 pdrs. Torpedo tubes: 1.	210
Battleships.	Oden .	1898	3,700	16.5	Guns: 2—10 inch, 42 cal.; 6—4.7 inch, 47 cal.; 10—3 pdrs.; 2—1 pdrs. Torpedo tubes: 1.	210
	Svea . Göta . Thule .	Reconstructed 1900-4	3,300	15	Guns: 1—8.2 inch, 45 cal.; 7—6 inch, 45 cal.; 11—6 pdrs.; 2—1 pdrs *Svea* and *Göta*: 1 tube.	237

SWEDEN.

		Completed	Displacement (tons)	Speed (knots)	Armament	Complement
Monitors.	John Ericsson .	Reconstructed 1898-1903	1,500	8	Guns: 2—6 inch; 2—6 pdrs.; 2 Machine.	80
	Thordön . .				Guns: 2—4.7 inch; 8—3 pdrs	
	Tirfing . .				Guns: 2—4.7 inch; 8—6 pdrs.	
	Berserk, Björn, Folke, Gerda, Hildur, Sölve, Ulf . . .	Reconstructed 1897 to 1909	460	8	Guns: 1—4.7 inch; 2—6 pdrs.	48
Armoured Cruiser.	Fylgia . .	1906	4,800	21.5	Guns: 8—6 inch, 45 cal.; 14—6 pdrs. Torpedo tubes (18 inch): 2.	322
Torpedo Cruisers.	Ornen, Jacob Bagge, Claes Horn . .	1896 to 1898	850		Guns: 2—4.7 inch, 45 cal.; 4—6 pdrs. Torpedo tubes: 1.	99
	Psilander, Clas Uggla	1899	850	20		

SWEDEN.

DESTROYERS.

6 *Wale class:* Wale, Ragnar, Sigurd, Vidar, Hugin, Munin. Speed: 31 kts. Armament: 2—12 pdrs.; 4—6 pdrs. Complement: 69.

1 Magne. Speed: 30·7 kts. Armament: 6—6 pdrs. Complement: 67.

1 Mode. Speed: 31 kts. Armament: 6—6 pdrs. Complement: 67.

TORPEDO BOATS.

17 *Plejad class:* Astrea, Castor, Iris, Plejad, Pollux, Thetis, Spica, Vega, Vesta, Argo, Antares, Arcturus, Altair, Polaris, Perseus, Regulus, Rigel. Armament: (except for Plejad, Castor and Pollux, which have 2—3 pdrs.), 2.6 pdrs.

2 *Gondul class:* Gondul, Gudur. Armament: 2 Machine guns.

12 *Blixt Class*—Blixt, Bris, Kapella, Komet, Meteor, Mira, Orion, Orkan, Stjerna, Sirius, Virgo, Vind. Speed: 23·20 kts. Armament: 2—3 pdrs

SUBMARINES.

3 New (R).

1 Hvalen (F). Speed: ~~~~ kts.

1 Hajen (D). Speed: $\frac{9}{7}$·5 kts. Complement: 6

URUGUAY.

		Displacement (tons	Speed (knots)	Armament
Light Cruiser . . .	Uruguay . .	1,400	23	Guns: 2—4.7 inch; 4—12 pdrs.; 6—1 pdrs. 2—18 inch tubes.
Gunboat	18 De Julio . .	678		Guns: 4 and 2 Mach

VENEZUELA.

	Maresal Sucre . .	1,030	14	Guns: 2—4 inch, 40 cal.; 2—6 pdrs.; 6—3 pdrs.; 2—1 pdrs.
	Bolivar . . .	571	10	Guns: 6—6 pdrs.
	Miranda . . .	200	12	Guns: 2—6 pdrs.; 1 Machine.
	Restaurador . .	568	12	Guns: 1—12 pdrs.; 1—6 pdr.; 2 Mac.
	Zamora . . .	740	—	Guns: 2 Machine.
	Zumbador . .	351	—	Guns: 2—2 pdrs.
	23 De Mayo . .	130	—	Guns: 1 Machine.
Torpedo Boat . . . (Old Thornycroft boat.)	Margarita . . .	97	. .	Guns: 4—3 pdrs. Torpedo tubes (14 inch): 2.

www.ingramcontent.com/pod-product-compliance
Lightning Source LLC
LaVergne TN
LVHW071515220926
848486LV00010BA/376